Mastering the Five Tiers of Audit Competency

The Essence of Effective Auditing

D0165327

Internal Audit and IT Audit

Series Editor: Dan Swanson

**A Guide to the National Initiative
for Cybersecurity Education (NICE)
Cybersecurity Workforce Framework (2.0)**
Dan Shoemaker, Anne Kohnke, and Ken Sigler
ISBN 978-1-4987-3996-2

**A Practical Guide to Performing
Fraud Risk Assessments**
Mary Breslin
ISBN 978-1-4987-4251-1

**Corporate Defense and the Value
Preservation Imperative:
Bulletproof Your Corporate
Defense Program**
Sean Lyons
ISBN 978-1-4987-4228-3

Data Analytics for Internal Auditors
Richard E. Cascarino
ISBN 978-1-4987-3714-2

**Fighting Corruption in a
Global Marketplace:
How Culture, Geography, Language
and Economics Impact Audit and Fraud
Investigations around the World**
Mary Breslin
ISBN 978-1-4987-3733-3

**Investigations and the CAE:
The Design and Maintenance
of an Investigative Function
within Internal Audit**
Kevin L. Sisemore
ISBN 978-1-4987-4411-9

Internal Audit Practice from A to Z
Patrick Onwura Nzechukwu
ISBN 978-1-4987-4205-4

Leading the Internal Audit Function
Lynn Fountain
ISBN 978-1-4987-3042-6

**Mastering the Five Tiers of
Audit Competency:
The Essence of Effective Auditing**
Ann Butera
ISBN 978-1-4987-3849-1

Operational Assessment of IT
Steve Katzman
ISBN 978-1-4987-3768-5

**Operational Auditing:
Principles and Techniques
for a Changing World**
Hernan Murdock
ISBN 978-1-4987-4639-7

**Securing an IT Organization
through Governance,
Risk Management, and Audit**
Ken E. Sigler and James L. Rainey, III
ISBN 978-1-4987-3731-9

**Security and Auditing of Smart Devices:
Managing Proliferation of
Confidential Data on Corporate
and BYOD Devices**
Sajay Rai and Philip Chuckwuma
ISBN 9781498738835

**Software Quality Assurance:
Integrating Testing, Security,
and Audit**
Abu Sayed Mahfuz
ISBN 978-1-4987-3553-7

**The Complete Guide to
Cybersecurity Risks and Controls**
Anne Kohnke, Dan Shoemaker,
and Ken E. Sigler
ISBN 978-1-4987-4054-8

Tracking the Digital Footprint of Breaches
James Bone
ISBN 978-1-4987-4981-7

Mastering the Five Tiers of Audit Competency

The Essence of Effective Auditing

Ann Butera, CRP

The Whole Person Project, Inc.,
Elmont, New York, USA

CRC Press
Taylor & Francis Group
Boca Raton   London   New York

CRC Press is an imprint of the
Taylor & Francis Group, an **informa** business
AN AUERBACH BOOK

CRC Press
Taylor & Francis Group
6000 Broken Sound Parkway NW, Suite 300
Boca Raton, FL 33487-2742

© 2016 by Taylor & Francis Group, LLC
CRC Press is an imprint of Taylor & Francis Group, an Informa business

No claim to original U.S. Government works

Printed on acid-free paper
Version Date: 20160104

International Standard Book Number-13: 978-1-4987-3849-1 (Paperback)

Library of Congress Cataloging-in-Publication Data

Names: Butera, Ann, author.
Title: Mastering the five tiers of audit competency : the essence of
effective auditing / Ann Butera.
Description: Boca Raton, FL : CRC Press, 2016. | Series: Internal audit and
IT audit ; 7 | Includes bibliographical references and index.
Identifiers: LCCN 2015045132 | ISBN 9781498738491
Subjects: LCSH: Auditing, Internal. | Auditing.
Classification: LCC HF5668.25 .B866 2016 | DDC 657/.45--dc23
LC record available at http://lccn.loc.gov/2015045132

Visit the Taylor & Francis Web site at
http://www.taylorandfrancis.com

and the CRC Press Web site at
http://www.crcpress.com

Contents

Preface

If you're reading this, you value effectiveness and efficiency. You like learning shortcuts that get results. You value time-tested approaches because you have better things to do than reinvent the wheel. You (and your boss) may be very happy with your performance, but you know that in this competitive work environment, treading water and living off past successes is a recipe for disaster. So, you are interested in ways to get your work done more efficiently without sacrificing quality. You are open to ideas that will make your work life easier.

Had operational auditing not become popular in the mid-1980s, I'm not sure that I'd be here, writing this book. But it did, and I am. Operational auditing opened up the discipline of reviewing the processes and approaches that people use to create value, provide support, and get results. This approach puts pressure on critical thinking and process and workflow analysis skills, in addition to the traditional risk and control evaluation skills.

The book's content is culled from over 30 years of experience consulting with hundreds of organizations and teaching thousands of internal and external auditors and operational, risk, and compliance professionals. Some of these folks had little or no audit experience; some were transferees from business units; some were college hires; some were career auditors. Despite the differences in experience, they all had one thing in common: they all needed to master techniques that would enable them to plan and complete audits effectively while managing their relationships with internal constituents and their bosses.

A funny thing happens when you teach: you learn. You learn how to explain things clearly to avoid confusing the participants. You learn how to communicate the same message in a variety of ways until the participants understand what you are trying to say.

Over the years, I started to see definite learning patterns emerge as I explained core auditing concepts and techniques. Certain questions would come up when specific topics were discussed. I honed my explanations and acquired an arsenal of time-tested approaches to accomplish the auditing activities. This book is my way of sharing these techniques outside of the classroom.

HOW TO GET THE MOST FROM THIS BOOK

Although this book may be read from cover to cover, it can also be used as a quick reference guide to spark ideas and solutions for situations you encounter at work. It is intended for auditors at all levels, including those just starting out. It contains an array of practical and time-tested techniques that foster efficiency and effectiveness at each stage in the audit. You can use it as a reliable resource when subject-matter experts or training guides are not readily available.

To be an effective auditor, you will need to develop five tiers of competency. This book explains that each tier comprises specific skills and behaviors. To the extent that you perform each of these behaviors effectively and efficiently, you increase the usefulness of your audit results and minimize audit risk. Regardless of the audit's complexity, the key question you need to ask yourself is, am I doing the right thing? Is my approach the right one or is there a more effective or efficient approach that I can use?

This book will spark ideas that will help you evaluate your performance in each of the five tiers. And it will give you the opportunity to develop a specific plan to enhance your competency. This will help you target specific skills you would like to develop, refresh, or reinforce.

I suggest that you start by defining your desired end state over the next 30 days. Do you want a promotion, recognition, less stress, or fewer review notes? Do you want to deliver tough news without engendering bad feelings, acquire data-analysis skills, run more efficient meetings, or write audit concerns more effectively? The more clearly you can define your objective, the easier it is to come up with an approach to achieve it. As you set your objective, keep in mind the comment by Frank Outlaw,* the late president of Bi-Lo, a supermarket chain.

Watch your thoughts, they become words;
watch your words, they become actions;
watch your actions, they become habits;
watch your habits, they become character;
watch your character, for it becomes your destiny.

Then think about how you want to use this book to help you achieve your goal. You can read the book from start to finish or you can simply focus on the areas of greatest interest and importance to you. Either way, think about what you do well that already contributes to your ability to achieve your goal. Consider the feedback you have received as well as the behaviors that others have noticed and admired— including the ones your boss has acknowledged. View these behaviors as your strengths. These are habits that you definitely want to retain and perhaps even take to the next level. At the end of each chapter, identify one or two techniques that will help you do this.

Then think about the competencies you want or need to develop that will help you achieve your objectives. As you read this book, identify specific techniques that will help you do this. Concentrate on practicing these new techniques every day for the next 30 days, because it generally takes this long to develop a new habit.

This book is divided into eight chapters. The first chapter, "How to Get the Most from This Book," sets the tone by defining the meaning of behavioral change and focusing on how to create and sustain it.

Chapters 2 through 5 address techniques and leading practices for completing an audit's planning, detailed risk and control assessment, testing, and reporting phases. For this book's purpose, the detailed risk and control assessment phase covers the audit activities related to

---

* 1977 May 18, *San Antonio Light,* "What They're Saying," Quote Page 7-B (NArch Page 28), Column 4, San Antonio, Texas (newspaper archive).

- Identifying and documenting risks and risk events
- Assessing the risk events
- Identifying and documenting controls
- Developing test plans to evaluate the controls

Chapters 6 through 8 address the interpersonal and social skills required to manage the audit team, the project itself, and the client relationship.

Since there's truth to the expression, "what gets measured, gets done"*, this book contains templates you can use to set performance goals for yourself and assess your progress toward achieving them. It explains how to make positive and sustained changes to the way you approach your work.

At the end of each chapter, you will see a summary of the key points and a brief quiz that will help you remember salient ideas and test your knowledge.

Also at the end of each chapter is the Performance Planning Worksheet (e.g., see next page). Its purpose is to help you achieve your goals over the next 30 days, which is the time needed to establish a new habit or routine†. Use the Performance Planning Worksheet to set your goals and remain focused on achieving them. You can also use it to identify the specific behaviors that you want to continue, stop, and start along with the observable behavior indicators you will use to assess your progress. If you are seriously interested in making change happen, then you need to do more than think and talk about it. You need to write it down. You need to see it in print, because writing clarifies thought. As you work toward your goal, remain patient and persevering.

This book will spark ideas that will enhance your performance, improve your working relationships with your team members and constituents, and make it easier to complete audits that improve your organization's risk management culture and practices. Wishing you all the best as you make positive change happen.

---

* Lord Kelvin in his May 3, 1883 lecture on "Electrical Units of Measurement" (*Popular Lectures*, Vol. 1, page 73).

† While it may take 30 days to start a new habit, staying away from a bad one is a lifetime effort, backed up by the fact that those well-worn synaptic pathways never go away.

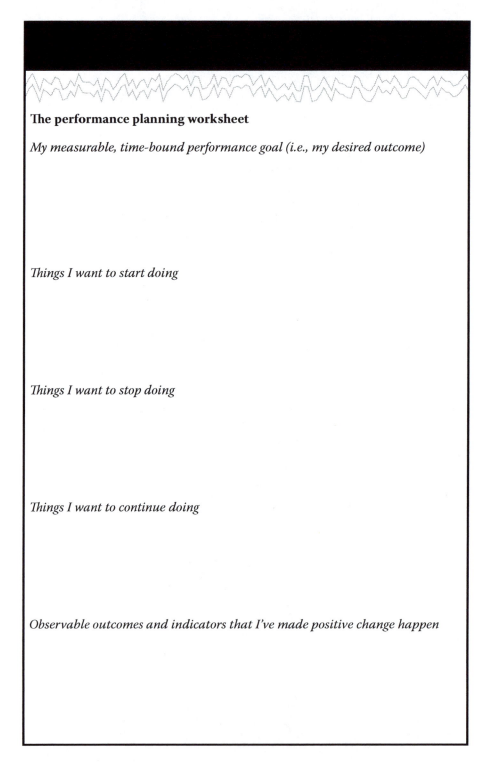

The performance planning worksheet

My measurable, time-bound performance goal (i.e., my desired outcome)

Things I want to start doing

Things I want to stop doing

Things I want to continue doing

Observable outcomes and indicators that I've made positive change happen

Acknowledgments

Writing this book was an endeavor that would never have been possible without the help and support from some very special people. Specifically, I am so grateful to

Dan Samson for suggesting that I write this book in the first place and for introducing me to Rich O'Hanley. His invitation sparked this endeavor.

Rich O'Hanley for agreeing to take on this project and for his advice and feedback. His experience has been so helpful.

Mariel Cruz for being the glue that held this project together and for doing the research and compilation that formed the core of the content. I have lost count of the number of times she reviewed the content and edited the materials. This book would not have happened without her effort and elbow grease. So many thanks for reading and, more importantly, applying the style guide!

Jesy Pijuan for developing the graphics and for "holding the fort" while Mariel and I were consumed by this book project.

Norma Bass, CPA for her real interest in the subject matter and her painstaking attention to the detail. I am most appreciative for her willingness to tap into her experience as a former chief audit executive and to spend countless hours discussing the nature of her observations with me just because we are friends.

Michael Hammond for his IT audit perspective, comprehensive comments, and attention to detail and consistency.

Ron Robertson for reminding me to stress data analytics and for valuing the interpersonal skills covered in this book as much as the technical ones.

Paul Flora, CPA, whose comments enhanced the content's clarity.

Amanda Singh for providing a user's perspective and helping to smooth the transitions.

Tim Santos for his continuous patience, understanding, and support, especially when writing this book precluded me from being able to play.

Those who have participated in the training programs over the years. Your comments, questions, and reactions have made me a better instructor and helped me hone my explanations and examples.

Author

Ann M. Butera is the founder and president of The Whole Person Project, Inc., an organizational development consulting firm that specializes in assisting companies to introduce, handle, and manage change. Her prior work experience as a department manager for a large retailer, a system liaison officer at a major money center bank, and an internal consultant at a bankcard processor, in conjunction with her experience as a business owner and consultant, enables her to develop customized approaches for continuous improvement and increased profitability in organizations of all sizes. She served as an audit committee chair for a financial services firm.

For over 25 years, Ann Butera has consulted with organizations of all sizes to provide managers, auditors, and compliance and risk professionals with the techniques needed to improve their risk management practices. She has worked specifically with audit, compliance, and risk management department managers to develop methodologies that enhance the consistency and quality of audit and review execution. Her consulting firm provides practical training in audit planning, risk assessment, control evaluation, and reporting, specializing in helping auditors communicate the risk and impact of their findings and concerns and to overcome constituent objections.

Ann Butera's expertise and experience in instructional design and her informative, entertaining, and thought-provoking presentation style have made her a sought-after lecturer for businesses, universities, nonprofit groups, and professional organizations at the international, national, regional, and local levels. As a lecturer and trainer, Ann Butera addresses critical business issues, providing her audiences with proven strategies that get results in a broad range of areas:

- Auditing and internal controls
- Business risk assessment
- Process mapping and flowcharting
- Measuring and tracking productivity
- Internal consulting skills
- Change management
- Oral and written communication skills
- Enterprise risk management
- Team-building techniques
- Audit sampling and testing
- Overcoming objections
- Leadership and managerial skills
- Audit methodology and quality improvement
- Performance appraisals
- Project management
- Interviewing, conflict resolution, and negotiation
- Operational auditing
- Strategic planning and corporate governance

She is a columnist for Protiviti's website, KnowledgeLeader, and has designed and delivered webinars for KnowledgeLeader and Corporate Executive Board. She has had numerous business articles published in many trade journals including *Bank Accounting and Auditing Magazine* and the *ABA Consumer Banking Digest*, and she has been quoted in *The New York Times* and *Working Woman* magazine, among other publications. She was a columnist for the Indiana Bankers Association's *Hoosier Banker* magazine.

She is a cofacilitator for the MIS Audit Leadership Institute and Chief Audit Executive (CAE) Masterclass and a frequent speaker at ISACA and Institute of Internal Auditors (IIA) chapter meetings worldwide. She was an instructor at the Bank Administration Institute (BAI) Bank School and the Southeastern School for Sales Leadership at Vanderbilt University. She was an adjunct professor in the Business Department of Long Island University and a core faculty member of the ABA's cable television show, *American Financial Skylink*. In addition, Ann Butera has hosted the radio show, *Business Forum*.

Ann Butera's professional accomplishments have merited recognition from national and local organizations. Beginning in 1984 and continuing on an almost annual basis thereafter, she has been cited in *Who's Who*. In 1990, Women on the Job honored her with the Business Achievement Award. A firm believer in volunteerism, she was elected president and chief volunteer officer of the Girl Scouts of Nassau County in 1991, serving as chair of the board until 1995. Ann serves on the Advisory Board of the Long Island Development Corporation. She is a member of the Institute of Internal Auditors, the American Society for Training and Development, and the World Future Society.

Ann Butera received her masters of business administration in organizational development from Adelphi University and is a summa cum laude graduate of Long Island University (C.W. Post Campus).

When she is not at work, she is either on the tennis court, in her garden, or at the theater.

1 How to Get the Most from This Book

CHAPTER CONTENTS AT A GLANCE

This chapter will

- Discuss the five tiers of competency that effective auditors require
- Explain the value of having habits
- Define three categories of change
- Describe the three behavioral stages of change management
- Explain how to make sustained change happen

The secret to change is to focus all of your energy, not on fighting the old, but on building the new.

Socrates

This book focuses on the behaviors you need to demonstrate and the habitual actions you need to take at each phase in an audit to manage the people relationships as well as the work itself. It contains practical techniques, advice, and tips that you can put to use immediately during an audit. It provides proven methods that will save you time, reduce your stress, and produce reliable, quality results. It will help you make positive changes and adopt productive work habits.

THE BOOK AT A GLANCE

This book is divided into eight chapters. The first five chapters are technical and are listed in the order in which the activities occur during an audit. From a career development perspective, this is not the order in which one would acquire these skills. If one were new to internal audit, one might spend three or four years involved in testing before acquiring sufficient experience to be tasked with planning an audit. Planning requires business acumen and auditor judgment; both are fueled by experience. Since the involvement of supporting staff in the audit typically occurs in the assessment phase discussed in Chapter 3 (especially in mid-sized and large audit departments), staff auditors aspiring to lead audits may want to read Chapter 2 to understand the considerations that affect audit management's planning and scoping decisions.

Chapters 6 through 8 tackle the personal and interpersonal skills needed to be an effective auditor and team member because technical expertise alone is not enough to produce relevant and useful results. In every business, management at all levels—especially at the board and executive ones—are responsible for driving business performance while managing risk to an acceptable level. Since some business situations are "safer" or less risky than others, management has to determine when it will accept the risk, transfer the responsibility for it (via insurance or agreement with other firms), or manage it by implementing controls. Additionally, even if management wanted to eliminate all risk, it could not for several reasons. First, some risks like interest rate risk, geopolitical risk, and acts of God originate outside the organization and cannot be prevented.

The role of internal audit is to function as the organizational mirror, reflecting back to management the condition and effectiveness of the control activities, monitoring, and policies they have implemented. Consequently, internal auditors need to influence management to take corrective action that will enhance the company's risk management culture and practices. Effective internal auditors require technical, interpersonal, and change management skills.

To help you be the best auditor you can, this book's underlying theme is how to create positive interpersonal and organizational change. By managing your behavior and adopting effective and efficient practices at each step in the audit, your efforts will contribute to your organization's risk management practices.

Read Chapter 1, "How to Get the Most from This Book," if you want to understand the Five Tier Auditor Competency Model and how to benefit from its use. It also describes your role in personal and organizational change and how you can facilitate change.

Chapter 2, "Techniques for Planning Useful Audits," will help you develop or refine the approach needed to complete the planning phase, particularly when setting the scope.

Focus on Chapter 3, "Techniques for Detailed Risk and Control Assessment," if you want to understand rapidly and accurately the nature and vulnerabilities of the process or entity under review. It will describe different approaches you can use to identify the risks in the process or entity under review. It describes the attributes that make up an effective control and suggests questions to ask to determine if a control is well designed.

Chapter 4, "Testing and Sampling Techniques," provides ideas that will help you develop and conduct effective audit tests and select the right sample size when full population testing is not possible. It also provides guidance for analyzing test results to determine their meaning.

In auditing, it's not enough to reach the right conclusion; one has to be able to produce useful, relevant written support. Consequently, Chapter 5, "Documentation and Issue Development: The Building Blocks for Effective Audit Reports," describes tips and techniques that will help you prepare effective, accurate documentation at each step in the audit.

Chapter 6, "Core Competencies You Need as an Auditor," explains guidelines for developing executive presence and critical thinking abilities. It also includes tips for managing your time and the project.

If you work in a mid-sized or large audit department, Chapter 7, "Techniques for Managing the Audit Team," explains ways to build and lead an effective team as well as how to contribute as an effective team member.

At some point in your audit career, you will have to tell the managers of the process you are auditing that controls gaps exist and that the inherent risk is not being managed to an acceptable level. Chapter 8, "Techniques for Managing the Constituent Relationship," describes ways to deliver bad news without creating bad feelings.

OVERVIEW OF THE FIVE TIER AUDIT COMPETENCY MODEL

Unlike other jobs, to be an effective internal auditor you need to be able to tap into and use five distinct competency categories, which are depicted in the Five Tier Audit Competency Model (Figure 1.1). A competency is the ability to do something effectively. It is based on experience and the skills acquired from that experience. For example, the ability to negotiate effective contracts is a competency, as it is the ability to organize and deliver clear and persuasive messages to allocate scarce or fixed resources between at least two parties.

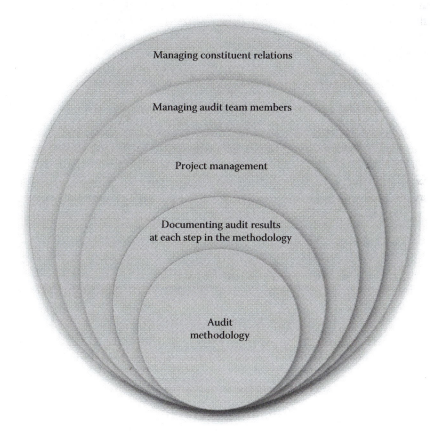

FIGURE 1.1 Five Tier Audit Competency Model.

Competencies are also referred to as skills. When you are able to demonstrate a skill with consistency and mastery—that is, you developed your skill through training and practice to an expert level—you would be described as being proficient.

To be an effective auditor, you will need to develop five tiers of competency. Each tier comprises specific skills and behaviors. To the extent that you perform each of these behaviors effectively and efficiently, you increase the usefulness of your audit results and minimize audit risk, that is, the risk that an audit does not identify an existing weakness and concludes that the controls are operating as intended when they are not. Regardless of the audit's complexity, the key question you need to ask yourself is, am I doing the right thing? Is my approach the right one or is there a more effective or efficient approach that I can use?

TIER 1: AUDIT METHODOLOGY

The first tier of the Five Tier Audit Competency Model, which forms the foundation for the other four, involves the mastery of the skills needed to execute an audit. If you don't have this competency, your career in audit will be very short. Some of the Tier 1 competencies are process analysis, risk identification, analysis and assessment, control identification and evaluation, test design and performance, and issue development. This tier encompasses the ability to demonstrate knowledge about sound auditing principles and professional skepticism.

TIER 2: DOCUMENTING AUDIT RESULTS AT EACH STEP IN THE METHODOLOGY

The model's second tier focuses on the documentation of the results of your thinking at each stage in an audit. The audience for each of these documents is not the same, and this difference in audience informational needs drives differences in writing styles. Following are some examples of the different types of informational needs your writing needs to address during an audit:

- Developing planning memos that describe the audit's scope, objectives, and resource requirements and include the rationale for these decisions
- Documenting the risk events and their consequences
- Documenting controls in narratives so that the effectiveness of the control design can be determined
- Describing test steps and the results of conducting them
- Compiling the report—the tangible product created at the end of a review or audit—that is read by an executive audience

Tier 2 competencies include the ability to

- Recognize and follow writing formats and styles at different points in the audit
- Write grammatically correct sentences that are grouped in paragraphs
- Satisfy the informational needs of a broad array of readers

- Produce stand-alone documentation that enables someone else to reperform the work and reach the conclusion you did

It is possible to possess the competencies in Tier 1; that is, an auditor may be able to think clearly and accurately when applying auditing principles and methodology but may not be able to document these ideas comprehensively and in a manner that would enable another auditor (or individual) to reperform the work, that is, demonstrate Tier 2 competency. Conversely, it is possible to possess strong writing skills yet lack the ability to apply auditing principles and methodology; that is, the documentation is organized and grammatically correct but omits critical or pertinent information or records the wrong results and conclusions.

TIER 3: PROJECT MANAGEMENT

The model's third tier relates to project management competencies. Project management is more than a skill. It is a profession that has its own designation from the Project Management Institute, Inc.: the Project Manager Professional (PMP)®. These are people who manage projects for a living—and these projects have nothing to do with internal audit. Tier 3 encompasses the skills associated with project planning, scheduling, time management, contingency planning, and estimating time and human capital resources.

Everyone assigned to the audit needs to be able to demonstrate project management competency—not just the project leader and audit managers. Staff auditors need to be able to estimate time requirements and manage their time to deliver useful results on schedule. Everyone needs to communicate accurate and comprehensive status reports.

TIER 4: MANAGING AUDIT TEAM MEMBERS

This includes the ability to

- Develop work assignments that optimize the talent of the audit team
- Delegate assignments to others
- Deliver useful and timely performance feedback and coaching
- Transform the auditors assigned to the review from a collection of professionals into a functioning team
- Develop the team members' competencies

TIER 5: MANAGING CONSTITUENT RELATIONS

The fifth and final tier focuses on the competencies associated with constituent relationship management. The scope of the competencies associated with this tier includes

- Acquiring technical and industry knowledge
- Demonstrating business acumen

- Dealing with critical conversations
- Delivering bad news without engendering bad feelings
- Communicating useful status information
- Negotiating useful corrective action plans

Given the array of competencies required to complete useful and value-added audits, you need to make sure that you are using repeatable and sound mental models and engaging in productive habits. The graphic at the beginning of each chapter will tell you in which tier of audit competency the content falls. This will allow you to focus on techniques and strategies targeted to a specific set of competencies. In my experience, auditors who master all or most of these competencies are more effective, become "go-to" people, and enjoy their jobs more.

SOME DEFINITIONS BEFORE WE START

Change: The act or process of becoming different. Change implies any variation whatever that affects something either essentially or superficially.

Change agent: A catalyst; someone who facilitates the change process. Generally, the change agent remains unaffected by the change; that is, he or she is an objective party that focuses on helping others adapt to achieve the desired outcome.

Future shock syndrome: This syndrome, identified by Alvin Toffler in his 1970 book *Future Shock*, occurs when the rate of change exceeds an individual's ability to assimilate it. This syndrome is characterized by immobilization; that is, individuals react as though they were deer caught at night in a car's headlight glare.

Managed change: A structured approach to the introduction and implementation of change, including ways in which internal or external experts and resources help organizations cope with resistance and other difficulties inherent to making change a reality.

Transformation: A major change in form, nature, or function.

CHANGE MANAGEMENT

HOW TO EFFECT PERSONAL CHANGE

Think about possible changes you could make that would position you to achieve your goal. What behaviors should you start to demonstrate on a regular basis? What habits should you adopt that will make your goal a reality? Record on the performance planning worksheet that will appear at the end of each chapter the one or two techniques that you want to start practicing on a regular basis.

Then, consider the current hindrances to your goal achievement. These are the current habits that you need to stop and replace with more productive actions because it is so difficult to simply go "cold turkey" to eliminate unwanted behaviors. If you don't believe me, talk to reformed smokers and others who have made a sustained and positive change in their lives. It's easier to substitute a behavior than to stop doing something cold turkey.

Once you've identified the behaviors that you want to continue, start, or stop and replace, prioritize them. If you want sustained results, you can only work on one or two changes in behavior at a time. Consequently, you need to focus on identifying the behavioral change that will generate the most leveraged result. What change will give you the biggest impact for your effort? Which change, once it has become part of your daily habits, will enable you to achieve your goal?

Then implement the new behavior, practicing it every day—even several times a day—until you own it. Now initially, you may forget to practice the new behavior. In essence, you may "fall off the wagon." That's to be expected when implementing anything new. When this happens, just get right back to practicing the new behavior as if the oversight never occurred.

THE VALUE OF FOCUSING ON BEHAVIOR

The emphasis on behaviors and habits is not an accident; it's quite deliberate. This behavioral focus is the essence of my work as an organizational development specialist. I help people create and sustain positive change in their work lives. This change may be precipitated by mergers, layoffs, acquisitions, new technology or systems, and other events that are outside the individual's direct control.

When most people meet me, they want to understand what I and my team do for a living. "Just what does organizational development mean?" they ask. It's an understandable question. The field of organizational development came into existence in the mid-1950s as a blend of sociology, psychology, and motivational theory applied to individuals in organizational settings.

As organizational development specialists, my team and I work with managers to define what good performance looks like; that is, what are the observable, measurable attributes that are present when work performance is effective and people are meeting their goals? We then study the actions of the good performers and we consider their backgrounds, education, and training. We also study the behavior of the poor performers or underperformers to understand and identify their actions. Essentially, we're looking for observable, measurable behavioral patterns. We then analyze these patterns and deconstruct them into their component parts and competencies so we can use this information to define what "good" looks like for a particular organization or situation. Then we use this information to develop recruitment strategies so that our clients can hire more good performers. We document the approaches that the good performers use to help our clients define sound, consistent, repeatable workflows and processes. We also develop training programs to teach others the skills, approaches, competencies, and habitual behaviors that they need to produce effective and consistent results.

Since this approach can sound esoteric, let me give you an example. One of our clients, a global finance department, wanted to build its bench strength and develop a succession plan. Management wanted to identify the attributes of high-potential people so they could identify those individuals who were already on the payroll, develop them in a concerted manner, and make them part of the organization's global succession plan.

At the start of this project, we met with the leaders of this department and asked them to define the characteristics of people who had successful careers within the organization. We organized their answers into four categories:

- Technical knowledge (which included things like experience in applying General Accepted Accounting Principles (GAAP) principles, experience in applying internal controls, merger and acquisition experience, international accounting experience)
- Business acumen, knowledge, and management (which included things like financial correlation skills, root-cause analysis ability, project management skills)
- Leadership and people management (which included things like experience in managing clerical staff and professionals, experience in leading major cross functional projects, the ability to garner support for new ideas, the ability to initiate and facilitate change)
- Other factors (which included things like being multilingual and an ability to relocate)

Each of these factors was analyzed to determine the underlying behavior or competency that acted as a success factor. For example, planning and organizational competency was a success factor for strategic planning and profit planning. The success characteristics and factors were recorded on a form that was used to rate each member of the global finance team. The managers used these results to craft career development plans for each finance team member. By keeping the focus on observable behaviors, the rating process was consistent across locations and was relatively easy to accomplish.

While the rating process to identify team members with high potential may be relatively easy to accomplish, getting the team members to change their behaviors is not ... unless the team member wants to change.

As auditors, we need to be able to promote change in our own lives so we can continue to grow and contribute to our organizations in a meaningful way. We also need to promote change within our constituents'* organizations to enable them to improve their risk management culture and practices.

THREE CATEGORIES OF CHANGE

Now that we have the terms defined, let's consider the types of change we encounter. Basically, there are three categories:

---

* Throughout this book, the term *constituents* is used to describe the people who are responsible for the audited area. According to Merriam-Webster, a "constituent" is "a member of a constituency or one of the parts that form something." In practice, internal auditors need to meet the needs of several constituents: the board of directors, executive management, and operating management. I considered using the traditional term, *constituent*, but rejected it because in English, the ending "ee" is typically the recipient of an action, that is, something is done to them. Since the internal audit process involves open communication between the auditor and business management, this term didn't seem appropriate. I considered using the term *business partner*, *client*, or *customer*, but rejected all of them because they were inaccurate. Internal auditors are not in a business partnership; that is, they don't share the profits or losses directly as partners would. The people who are audited are not customers or clients because they do not directly pay for internal audit's services and they cannot terminate the internal auditors' services.

Micro Change

This change affects us directly as individuals and includes changes that affect our family and friends. Examples of this type of change include

- You are getting married.
- You are expecting a child.
- You or someone in your family loses a job.
- Someone in your family dies.
- You buy a new home.
- Your child's local daycare center closes or your child's local daycare service provider can no longer work with you.
- Your gym closes.

Often our coworkers and boss may not know that any of these changes are happening unless we want or need them to have this knowledge.

Organizational Change

This type of change directly affects our work life, career, and employer.
Examples of this type of change are

- You get a new boss.
- Your department and function have been reorganized.
- Your company has been acquired.
- Your industry—for example, print media—is experiencing volatility.

Macro Change

This change directly affects large constituencies such as entire sectors or industries outside an individual's immediate world.
Examples of this type of change are

- The USSR ceased to exist.
- An election has occurred and a new political party is in power—unless you work for the government, in which case this type of change would be categorized as organizational.

These definitions describe three categories of change. Regardless of the category, we as individuals respond to change in predictable ways.

How Individuals React to Change

The typical reaction to change can be divided into three phases:

Phase 1: Relinquishing the Old Reality

For some, this phase is characterized by initial anticipation concerning the expected outcome of the change and possible unfounded optimism; that is, the change outcome will be so wonderful that it will solve all of our problems, including acne, famine, and floods. Individuals with this type of outlook look forward to the change and view it positively—at least at first—without any rational basis for doing so.

As an example, imagine that your audit department recently adopted or changed its audit software. When this project was originally announced, anyone who was elated to hear the news and anticipated a streamlined, problem-free audit methodology would be characterized as displaying unfounded optimism. These folks envisioned a blissful world in which all work papers and documentation would be in one place. They viewed the new software as the panacea that would correct all the flaws in the current methodology and make sharing information and progress tracking easier. Reality sets in when these same folks discover during the training period that the fields are difficult to fill in, audit activities are out of the accustomed order, and they must learn new codes. In the later stages of the software implementation, they might experience feelings of denial, blame, and anger at the prospect of letting go of their existing reality.

Phase 2: Entering the Transition Zone

This phase is defined as the interim period between what is gone and what is not yet in existence. A person can remain in the transition zone for an indefinite amount of time. Some people never come out.

Initially, people experience disorientation: "Everything has fallen apart"; "I knew what I was, but I don't know what I'll be." In later stages, particularly if a person is to move out of the transition zone, he or she becomes receptive to the opportunity to create new beginnings.

Have you ever seen the 1999 movie *Office Space*? If you haven't seen it, watch it while thinking about how people accept and effect change. Specifically, pay attention to the character Milton the Stapler Guy. Initech, his employer, was in the process of reorganizing—everything. This company even switched stapler brands, something Milton couldn't and didn't accept. Initech had fired Milton five years earlier, but he was so immaterial to them that they didn't notice he never left and kept paying him. His boss moved his desk to increasingly worse areas, ultimately relegating him to the rat-infested basement. The boss even confiscated his red Swingline stapler, Milton's only ray of sunshine. (Who doesn't appreciate a functioning stapler?) Surely, most of us would have left at the first instance of disrespect, or we would at least acknowledge that our manager's treatment was grossly inappropriate. Milton was unable to get past the initial disorientation stage of Phase 2. He was bewildered by his treatment but could not and did not accept that he was no longer employed. Eventually his simmering rage boiled over and he burned down the office.

How could Milton have changed his situation? He could have acquired a new skill to make himself valuable to the company again, or he could simply have accepted his termination and looked into becoming a Swingline salesman.

Phase 3: Accepting the New Reality

The behaviors that characterize this stage include the individual's testing and receptivity to new ideas. During this phase, the person begins to identify with the new vision and identifies the tools, knowledge, and skills required to perform in the new reality. Once initial skills, tools, and knowledge are identified, the individual begins to build the needed competencies to function and succeed in the new reality.

SPECIFIC WAYS TO INCREASE ACCEPTANCE OF CHANGE

There are several things that you can do to make it easier for you to accept and adapt to changes in your work, organization, and life. You can also adapt these approaches to make it easier for others to deal with and accept change.

Let's focus first on the things you can do during Phase 1: Relinquishing the old reality. While it is tempting to ignore the need to change, this will only prolong the process and put you at risk of missing opportunities. You should concentrate on articulating the new vision as clearly as possible to make sure that you understand the nature of the changes that will occur. There's one way to demonstrate that you understand something: if you can explain something to someone else and they understand what you have said.

Once you understand the nature of the planned changes, make sure that you clearly understand the ramifications of the change you face. What will you need to do differently as a result of this change? To help you do this, consider and identify (preferably in writing—because writing will clarify your thoughts) as specifically as possible the future impact of this new vision on you and your job position as well as your concerns, fears, and questions.

Articulate your concerns and fears to your manager, asking for feedback that addresses these as completely as possible—but understand that your manager will probably not have all the answers at this point in the change process.

Consider your manager's responses as they address or ameliorate your concerns. Once you acknowledge that the impending change is real, you can take the following actions to make it easier to get through Phase 2: Entering the Transition Zone.

Target the areas where your feelings of greatest loss exist. Address these feelings of loss and assist the acceptance of change by creating possible new beginnings. Perhaps some of your work friends left—make plans to stay in touch, and make an effort to get to know new colleagues. They are experiencing change too, albeit from a different perspective. Pinpoint the benefits of a new team and engage in new team-building activities.

New colleagues can cause perceived or real loss of turf. Avoid becoming territorial—focus on the results the shift of responsibilities will bring. Is it possible you were taking on too much before and now there is less on your plate? Or is management playing to your strengths in giving you new or focused responsibilities? What are some ways you can transition your existing skill proficiency?

Generate ideas concerning how your new role relates to the new organizational goal or vision.

Schedule another discussion with your manager to learn how you will or can fit in to the new vision. Ask how you can leverage the planned changes into new career opportunities to explore.

During this phase, you need to keep an open mind, consider the competencies you enjoy using and the ones you'd like to acquire. Then, brainstorm the various ways you can put your existing (and future) skills to use to contribute to the new reality.

Once you accept the change, you still need to demonstrate that you accept the new reality. During Phase 3, you need to exhibit consistency and reliability in your performance so that the other members of your team and in your organization have confidence in you.

WHY CHANGE IS NEVER EASY

Although Darwin's theory of evolution, loosely paraphrased as "adapt or die," has been discussed for years, the fact is that change is almost antithetical to ordinary human behavior. People change only when the benefit derived from the change exceeds the benefit derived from the status quo. Once a chaotic or problematic condition is resolved, the human tendency is to create order and then rest, having regained a sense of equilibrium. Consequently, it is a rare individual that seeks unending development characterized by ceaseless change. Most people opt instead for the comfort and predictability of "tried and true" routines.

Inherently, change involves risk. It is a deliberate migration away from the comfort of known routines and approaches without the certitude that the effects of change will occur as desired. On an individual level, change requires the creation and adoption of new and unfamiliar behaviors. One has to consciously accept that change is necessary and can result in a condition that is better than the current one.

Making change happen means putting an end to habits and routines. A habit, by definition, is a behavioral pattern that is performed automatically when triggered by a cue. Habits are time-saving routines that enable us to make decisions effortlessly. For example, most people have a "morning routine" that gets them awakened from their sleep, cleaned up, and on their way to work. Once at work, most people have another morning routine, which enables them to organize their time. Habits prevent people from having to make countless, time-consuming decisions like

- Should I wake up on the right or the left side of the bed this morning?
- Should I have coffee or tea as my morning breakfast beverage?
- Should I put my shirt or pants on first?
- Should I put on my right or left shoe first?

I could go on, but I think you get the idea. Habits have their value. Some habits are very productive; usually these are the ones that contribute or have contributed to successful outcomes. But, like anything that's overused, habits can be problematic. When you want to create a change, like getting to work earlier, exercising more, or being more organized at work, you need to create new behavioral patterns to replace the ones that you want to change. Now, you might be thinking, "I don't have any behavioral patterns like the ones you've described. I'm a 'free spirit', open to whatever happens each day. I let the day's events drive what I do." Then, I would say, being that free spirit waiting to react to a situation IS your behavioral pattern.

Sometimes change is foisted on you and it's not something that you want to happen. I am alluding to changes triggered by external events like mergers, acquisitions, bankruptcies, natural disasters, and economic disasters. In these situations, we are not the willing. In fact, our reaction to this type of change parallels the human reaction to death and dying characterized by Elizabeth Kübler-Moss in her 1969 book *On Death and Dying*: denial and isolation, anger, bargaining, depression, and acceptance.

For sustained behavioral change to occur, two things need to happen. First, you need to become aware that change is needed, that is, that your current style or actions are either unproductive or counterproductive to your goal achievement. Secondly, once you are aware of the need for change (and some people never gain this awareness), you need to decide on the nature of the change that needs to occur. While deciding on the change, you may have to experiment with a few different behaviors until the desired results are achieved. This can take months and requires a great deal of perseverance and personal discipline.

So, if you want to make a positive change, focus on the result you want to achieve and determine the behavioral change you need to make to reach this goal. Essentially, work backwards from your desired end state to develop your own action plans for change. Then, focus on implementing the behavioral change for the next 30 days, after which time the new behavior will have become a habit or routine.

Think carefully about the behavioral change that you need to make. By targeting the right behaviors—the ones that are integral to your performance—you may only need to focus on one or two key changes. Let me give you an example.

WHY TARGETING THE RIGHT BEHAVIORAL CHANGE IS SO IMPORTANT

Muffy was chronically late to work. Sometimes, she was 5 minutes late; on other days she could be 30 minutes late. She always had a reason for her tardiness; there was an accident or too much traffic, or a power outage had caused her alarm to malfunction. These things happen; they just happened to Muffy more frequently than to her team members. From Muffy's perspective, her lateness was inconsequential and certainly not a performance issue. She produced good work, completed her assignments by their due date, and stayed late when necessary to accomplish her projects. However, Muffy realized that her lateness was not unnoticed by her boss, who was a time-management fanatic. Muffy sensed that if she didn't proactively address this situation with her boss, she would have a performance problem. So, Muffy scheduled a one-on-one meeting, during which she acknowledged that she had difficulty getting to work by 9:00 am because of other responsibilities. She asked her boss whether she could change her work schedule and start at 9:30 am instead of 9:00 am and stay 30 minutes later at the end of each day. Her boss agreed, and the schedule change went into effect the very next day. But, instead of arriving to work at 9:30 am, Muffy showed up at 9:45 am!

Why didn't the schedule change work for Muffy? She requested it; it wasn't imposed by her boss. The schedule change didn't work because it was not the root cause of Muffy's tardiness.

Muffy didn't target the right behavior. She mistakenly thought that by changing her start time, she would be able to arrive on time to work. However, the root cause of the problem wasn't her work schedule; it stemmed from work–life balance issues—and her life was out of balance. Consequently, family obligations kept her up until the early morning hours, leaving her sleep deprived and incapable of waking up on time to make it to work. Once she set boundaries between her work and personal lives and learned to say no at the right times, her life regained equilibrium. Making this change wasn't easy for Muffy. It took time and concerted effort.

I wanted to make a behavioral change when I first started my practice in the early 1980s. I spent a lot of time on the phone speaking to people in various parts of the world. If the person I was trying to reach was not in, I would leave a message. In that message, I key-spelled my name and left my phone number to make it easier for them to call me back. I would spell my name: B for boy, U for umbrella, T for Thomas, ERA. Now, as a lifelong resident of New York, I have a distinctive regional accent. I realized that my accent affected the way I pronounce certain words—like the word "for" as an example. When I say it, it sounds like "fur." So, what my listeners heard me say was: B-fur-boy, U-fur-umbrella, T-fur-Thomas, ERA. And of course, if I was in a hurry when I left the message, I spoke faster and the mispronunciation was more obvious.

I decided to replace "for" with "as in" because I knew that this would be easier for me to pronounce without an accent. Whenever I would leave a message, I would say B as in boy, U as in umbrella, T as in Thomas, ERA. Initially, it was tough to make this change. I kept reverting to my old ways. But I was determined, and eventually the change occurred. More importantly, the change has lasted. I replaced an old behavior with one that I still use today. Essentially, that's what this book is about: replacing unproductive behaviors or habits with productive ones.

While change is never easy, it is significantly easier if you are the one that has initiated the change instead of the one expected to make the changes. For example, if you are the one with the vision of how the risk-based audit methodology should work, you promote the approach staunchly. In fact, you may have trouble understanding how anyone could be opposed to the approach. To you, this methodology's benefits are self-evident and no explanation is really necessary. Conversely, if you are the one expected to use the new risk-based tools, software, and reporting formats, you might not even appreciate why the change is needed, let alone yearn to master the new method, techniques, software, and documentation requirements.

WHY CHANGE IS EVEN MORE DIFFICULT IN AUDIT DEPARTMENTS

Audit departments face unique challenges because they attract individuals with particular behavioral preferences. These predilections are directly observable behavioral styles, for example, patterns of speech, action, and interaction. Thankfully, unlike genetic immutable conditions, individuals can change their behavioral styles—once they realize how their propensities are hurting them.

During the past 20 years, I have assessed and profiled thousands of auditors ranging from staff auditors to chief audit executives. This assessment enables individuals to categorize their own behavioral preferences. The behavioral styles are summarized in Figure 1.2. As you review this information, keep in mind that these are typologies and that the percentages reflect data collected from audit departments with various organizational cultures. Real people in all professions reflect aspects of all styles as well as secondary preferences, that is, backup styles. I discuss these in more detail in Chapter 8 on page 161.

People-pleasers	**Analyticals**
Likely to be warm, friendly, and helpful. They are concerned about how ideas, recommendations, and changes will impact themselves and others. They may look for help or consensus from others before making a decision.	Likely to speak in measured tones and be interested in the details. They may want all arguments proved logically, with figures to back up claims, before deciding.
Bottom-liners	**Vocalizers**
Likely to be direct and abrupt. They want to get to the point without "beating about the bush." If they see the benefit, they will make immediate decisions.	Likely to be effusive and talkative, with a constant flow of thoughts and ideas. It can be hard to keep up with them or get a word in edgewise! They have lots of intentions and can easily procrastinate so you have to pin them down to a commitment.

FIGURE 1.2   Summary of behavioral styles.

Clearly, people with a predilection for structure are attracted to internal auditing. On the surface, you might think that this behavioral preference would be ideal for audit. And, in general, it is. In order to appreciate the challenges this propensity for structure poses, you need to accept a fundamental premise: one's greatest strength is also one's greatest weakness. This paradox occurs because individuals tend to repeat behaviors that result in success and predictable outcomes. This leads to an overuse of one's strengths.

In audit, the ability to understand performance criteria, compare organizational or process results with this standard, and identify and report on variances is at the core of an auditor's critical competencies and is the hallmark of effective auditors. This same strength, when overused, can cause the auditor to become rigid and incapable of perceiving shades of gray in a business situation. When this predilection is overused, the auditor relies increasingly on checklists and performs less and less synthetic thinking. In extreme cases, the auditor's behavior can simulate little more than a walking internal control questionnaire. Audit scopes and objectives cease to reflect the unique aspects of constituent businesses. Even this auditor's reports can begin to reflect repetitive information.

Change is anathema to someone with a predilection for structure and repetitive routines because change is rife with uncertainty. Additionally, risk-based process auditing requires a great deal of synthetic thought, typically more than is required for a general controls review or a routine compliance audit. Auditors have to cull a great deal of information and understand a process that spans several departments and locations in order to determine what aspects are material. Decisions concerning what should be in and out of a process audit's scope require a certain level of auditor risk-taking. Let's face it; decisions regarding an audit's scope may be too narrow. Inherent in these decisions is the risk that some important aspect of the process will be overlooked. As if this audit risk were not enough, there is always the vulnerability that a recently audited area will "blow up" shortly after a satisfactory audit opinion is issued.

KEYS TO ACHIEVING SUSTAINED CHANGE

So, if you want to establish and maintain a new habit, keep the following points in mind:

- Possess a true desire to make the change a reality. You need to really want to change your current habit and introduce a new one. True desire happens when you realize that if you change your behavior, your future will be better than your present. Sometimes a cataclysmic event, like the death of a close family member or friend, losing one's job, or being overlooked for a promotion that you thought you deserved, will cause you to realize that you need to make a change.
- Commit to a strong belief that you will be successful in dealing with the change. If you don't believe that you will be successful, you won't.
- Create a plan for change or a strategy you'll follow to achieve results. This plan enables you to follow a series of steps to make the change occur.
- Acknowledge that setbacks are a natural part of the change process and actually can provide a motivation for making or accepting a change. View setbacks as mistakes you've been given the chance to correct.
- Celebrate progress—great or small—when it occurs. Feeling good about what you've accomplished to date will help to motivate you to continue until you've completely accepted the new reality.

While reading this book, you may realize that you have to make some minor or even major adjustments to the way you approach an audit. If change is hard for you, the tips in this book can help you accept it more easily.

CHAPTER SUMMARY

- Use the Five Tier Audit Competency Model to develop, refresh, or reinforce key audit behaviors.
- Start to effect change by defining your desired end state over the next 30 days.
- Understand your own and others' behavioral style to identify how change affects you.
- Target the right behaviors when planning change.
- Making change happen means putting an end to habits and routines—accept your own and others' setbacks and celebrate progress.

QUIZ

1. A competency is mostly based on
   a. Innate ability.
   b. Skills acquired through training.
   c. Skills developed from experience.
2. If you want sustained results, you should
   a. Work on all your desired changes at once.
   b. Work on changing one or two behaviors at a time.
   c. Hire a professional coach.
3. The three types of change are
   a. Macro, micro, and organizational.
   b. Situational, focused, and widespread.
   c. Personal, business, and foreign.
4. To achieve sustained change, you must target the right behaviors:
   a. True
   b. False
5. A strength can become a weakness when
   a. It is not used enough.
   b. It is overused.
   c. It clashes with others' abilities.

The performance planning worksheet

My measurable, time-bound performance goal (i.e., my desired outcome)

Things I want to start doing

Things I want to stop doing

Things I want to continue doing

Observable outcomes and indicators that I've made positive change happen

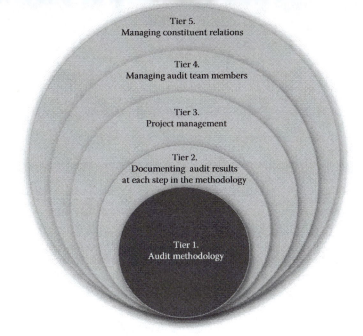

Tier 5.
Managing constituent relations

Tier 4.
Managing audit team members

Tier 3.
Project management

Tier 2.
Documenting audit results
at each step in the methodology

Tier 1.
Audit methodology

2 Techniques for Planning Useful Audits

CHAPTER CONTENTS AT A GLANCE

This chapter will

- Define the Critical Linkage™ technique and how to apply it during your reviews
- Describe tips to help you acquire a comprehensive understanding of the process under review
- Explain techniques that will help you identify, categorize, and analyze business processes
- Explain "funneling"—an interviewing approach to get the best and most information from others
- Outline six factors to consider when setting realistic scopes
- Estimate the time requirements

Luck is what happens when preparation meets opportunity.

Seneca

There's something simultaneously exciting and scary about planning an audit. On the one hand, you are starting with a "clean sheet of paper." You are in a position to define the audit's objectives, scope, and time frames. You determine whether special resources are needed. You decide how you want to divide up the work to be performed. Assuming you are not the only person assigned to this review, you assign the work to the staff, trying simultaneously to optimize their experience while developing their skills.

During the planning, you have a lot of information to collect and digest and many decisions to make. Of course, you also need to get audit management's approval for all of these decisions, which means that you need to be able to support your conclusions with facts. But the point is that you are directly setting the audit's scope and direction.

While all of this is going on, you need to establish a positive working relationship with the senior management of the area under review—and in some cases there's a disparity in their organizational rank and power compared with yours. Your relationship with these people will set the tone for the review. It will color your working relationship with the senior manager's direct reports and affect their cooperation and communication with you and your team.

On the other hand, the decisions you make during this planning phase shape the future of the review and can create audit risk if you decide to exclude an area that should have been included. Given that your boss will review and approve your decisions before you and your audit team get too heavily into the detailed risk and control assessment activity (assuming of course that you don't work in a one-person audit department), the likelihood of you omitting a key area to review is probably low. What is more likely, though, is that you and your team may not have sufficient subject-matter expertise, especially if the audit involves information technology (IT). At a minimum, you will always—admittedly, that's a big word—have less subject-matter knowledge than the manager of the area you are reviewing.

Before I get too far into this chapter, let's define some terms. For the purposes of this chapter, I define *planning* as the activities needed to acquire an understanding of the area under review, set the audit's objectives and scope, determine the resources required to complete the project, establish the time lines and deliverable due dates, and officially kick off the audit with the area's management. Depending on your department's methodology, some of the activities I've just described may be performed during a preliminary work phase (Box 2.1).

Regardless of what you call the audit phase during which you need to do these things, audit planning requires experience as well as the ability to

- Define and apply the Critical Linkage™ to an entity's objectives, risks, and controls
- Research information concerning the entity under review
- Use data-collection interviewing skills to get the most and best information from others
- Acquire a balanced understanding of the area under review
- Analyze processes to perceive patterns and identify vulnerabilities; understand their objectives; determine their inputs, outputs, and interdependencies; and articulate their value within the context of the rest of the organization

> ## BOX 2.1 THE DEFINITION OF PLANNING
>
> The activities needed to acquire an understanding of the area under review, set the audit's objectives and scope, determine the resources required to complete the project, establish the timelines and deliverable due dates, and officially kick off the audit with the area's management.

- Discern salient from irrelevant or inconsequential facts
- Think critically about the information to decide the best use of the time allotted to the audit

Whew! That's a lot of work. High levels of talent and energy need to be expended during planning. It's not by accident that the primary responsibility for planning audits is assigned to senior auditors. Let's look more closely at some of the competencies needed during planning, beginning with use of the Critical Linkage™.

DEFINITION AND BENEFITS OF THE CRITICAL LINKAGE™

The Critical Linkage™ is the relationship between an entity's corporate and functional objectives and major process steps (referred to collectively as Step 1—Objectives), the relevant inherent internal and external risks (Step 2—Risks), the design and operating effectiveness of the associated key controls (Controls), and, in cases where the controls are poorly designed or not working as intended and the amount of residual risk is excessive, the corrective action (Step 4–Action Plan) management commits to take. This Critical Linkage™ is based on the Committee of Sponsoring Organization's (COSO) Internal Control Framework and is a practical way to think about and approach risk-based process auditing.

When you apply the Critical Linkage™ to the entity you need to audit, you focus on the areas of greatest inherent risk and concentrate on evaluating the key controls that are intended to mitigate these risks. When you understand the linkage of an entity's objectives, risks, and controls, you are able to sustain meaningful conversations with the entity's management concerning operational vulnerabilities and acceptable residual risk levels. You are able to set audit objectives that optimize the time allocated for the audit while minimizing audit risk.

There's one other benefit that is especially important if you are a new audit project lead. When you apply the Critical Linkage™, you are able to ramp up and rapidly acquire an understanding of the area under review—even when it is totally unfamiliar to you!

APPLYING THE CRITICAL LINKAGE™

Let's look at the actions required at each step in the Critical Linkage™ so that you can apply it to your risk-based and performance audits as well as your preimplementation reviews (Figure 2.1).

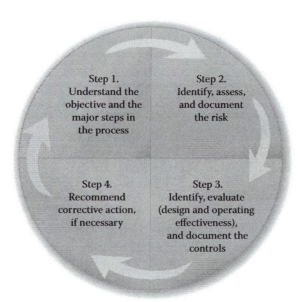

FIGURE 2.1   The Critical Linkage™.

Step 1: Objectives

At this point, you need to understand the business or activity objective(s), that is, what the area, entity, process, or project under review is trying to accomplish. This business or activity objective should also have

a. An *operational* component, that is, what needs to be done; what results need to be accomplished; the dollars to be saved, spent, or raised
b. A *reporting* component, that is, the accurate recording and reporting of the area's operational, compliance, financial, and strategic data
c. A *compliance* component, that is, all applicable regulations and policies will be upheld
d. A *strategic* component, that is, the daily activities are congruent with and contribute to the achievement of the organization's strategic direction.

As part of this step, you need to identify the major activities involved in achieving the business objective.

Note: The business objective may be strategic (e.g., new product development) or tactical (e.g., making trades, building servers). At this point, do not consider any controls—it's premature because you haven't yet considered the inherent risks.

As an example, you may want to ask:

• What are the written business or process objectives? (Don't be surprised when there are none. Many managers and their teams are so busy doing their work that they don't take time to write business objectives.)
• How do the business or process objectives encompass strategic, operational, financial, and compliance components?

- How do the business or process objectives relate to the overall organizational mission and strategic plan?

Step 2: Risks

As part of this step, you need to identify the inherent internal and external risk events, that is, those negative incidents that threaten the achievement of the business or process objective. As part of this step, analyze the risks to identify those that would have a higher impact and have a higher chance of occurring.

Identify the risk events by asking, "What could go wrong to prevent the achievement of the business objectives and the major steps needed to achieve this objective?" The inherent risk identification process begins with a firm understanding of the business objectives for the activity under review. At this time, you should also consider the nature of the organizational culture and internal environment—that is, the tone at the top—when assessing inherent risk.

As an example, you may want to ask

- What internal and external factors (or issues) could prevent the achievement of the goals of the activity?
- What would trigger or cause the risk event?
- How bad would the consequences of the risk event be?
- How likely is the risk event to occur?
- If this risk event occurred, how long could it last?
- How quickly could either the impact or news of this negative outcome spread?

We will get into this topic in more detail during Chapter 3.

Step 3: Controls

As part of this step, you need to identify and evaluate the existing control environmental factors, control activities, and management monitoring to determine how economically, efficiently, and effectively they address and manage the risks. During this step, you consider the control's design as well as its operating effectiveness. It is during this step that you will use inquiry, observation, and walk-throughs to evaluate the control design and then create test steps to determine whether the well-designed controls are operating as intended.

As an example, you may want to ask

- How are the inherent risks managed, prevented, detected, and corrected?
- How does management know that the work and the results are being performed according to plan?
- What steps are used to monitor results? How often? Who signs off on reviews?
- How is noncompliance with policies handled? To what extent are exceptions made based on an individual's position and background?
- Is there too much monitoring?

We will get into this topic in more detail in the next chapter.

Step 4: Corrective Action

Based on the results of Step 3 in the Critical Linkage™, management may need to develop an appropriate action plan to correct situations in which the controls are either nonexistent, poorly designed, not functioning as intended, or inefficient. Management's action plan specifies who will do what by when to resolve the situation. Effective action plans are SMART: specific, measurable, achievable, relevant, and timebound.

As an example, you may want to ask

- Who is responsible for doing what by when?
- What are the desired outcomes once the action plan is implemented?
- Will the action plan satisfactorily address the risk issue?

By using the Critical Linkage™, you establish an intellectual cohesion that begins with the audit entity's objective and ends with how effectively the entity's controls are mitigating the inherent risk. This thought process enables you to arrive at meaningful residual risk levels. It also enables you to talk convincingly to the entity's management about the effectiveness of the risk management practices they use. It enables you to audit unfamiliar areas with confidence.

Now, you may not be responsible for planning the audit. Yet, you can still use the Critical Linkage™. Let's assume that you are a staff member assigned to an audit, and the audit scope and objectives have already been documented and approved. Let's assume that you've been assigned to test the controls. While you could simply look at the audit program, review the test steps, pick your samples, and start testing, you will be in a much better position to conclude on your findings if you understand the Critical Linkage™ first. If you make the small investment in understanding the area's objectives and the purpose of the process or function to which you are assigned, you will appreciate how the inherent risks can occur. This knowledge, in turn, will make it easier for you to determine the key controls needed to mitigate the risk and evaluate the ones that management has put into practice. Once you understand each key control's purpose and intended operation, you will understand what a deviation means and how to express this in writing as an audit observation—all before you have picked your first item to test!

If you're a veteran auditor, someone who's audited the same areas for years, you can benefit from using the Critical Linkage™ as well. It provides a framework for data collection that will enable you to amass planning data more efficiently. It also enables you to articulate the inherent risks clearly and in terms that will be meaningful to the area's management as well as to any less experienced auditors assisting you in completing the review. Leveraging your understanding of the objectives and major steps in the process as well as the inherent risks will enable you to set meaningful audit scopes and objectives—ones that add value while minimizing the audit risk.

Let's focus a bit more on accomplishing the first step in the Critical Linkage™: gaining an understanding of the objectives and process under review.

An Example of the Critical Linkage™ in Action

A few years ago, I trained Gizmon (not his real name), a recent college graduate who had secured a spot on one of my client's audit teams. Gizmon's bachelor's degree was in economics, and he had never taken an auditing class as part of his undergraduate studies. As part of my client's new auditor orientation, Gizmon was required to participate in a risk-based audit methodology class that I had designed. This class emphasized the Critical Linkage™, and he was an engaged student. Three months later, I ran into Gizmon while I was in the United Kingdom teaching another class for this client. He was getting a cup of coffee at the same time my class was enjoying an afternoon break.

Gizmon was in the process of identifying, documenting, and evaluating the controls in a highly specialized and technical area. When I saw him, he was elated because he had just finished explaining his audit findings to one of the area's managers, and the manager had agreed with his observations and conclusions. He attributed his success to his use of the Critical Linkage™. When he learned that he was going to be assigned to this audit, he was a little nervous. He was aware that he knew nothing about the area under review. So, he read the audit planning documentation the audit's lead auditor prepared, visited the area's intranet site, and met with the lead auditor to get answers to his questions. From Gizmon's perspective, the Critical Linkage™ gave him the context he needed to acquire an understanding of the area under review. He then leveraged this understanding to identify the risks and controls. Using the Critical Linkage™ helped him acquire confidence, interact intelligently with the constituent, and point out a gap in the control environment.

AUDIT PLANNING TECHNIQUES

How to Understand the Entity under Review

An organization's audit universe comprises auditable entities. Some of these entities will consist of processes, for example, loan origination or payroll, and when this happens it is easier to determine the entity's boundaries and the audit's scope. More often, the audit entities are departments or functional areas responsible for all or part of several different processes, for example, finance, human resources, or IT. When this happens, it is important to articulate the entity's boundaries and describe the processes that constitute the entity prior to setting the audit's scope. If the boundaries are not accurate, key activities could be overlooked or the audit scope may be set too broadly and create scope creep.

The following techniques will help you gain an initial, broad-brush perspective of the area you need to audit:

Focus on the business area overall. Consider questions such as

- What does this business area or process do?
- Why does it do this?

- What are its overall business strategy, goals, and objectives?
- Where is it located?
- Who are the key upstream and downstream customers?
- Who are the key staff contacts?

Focus on the external environment. Consider questions such as

- What general economic factors affect this business area or process?
- What specific economic factors are affecting this business area or process currently?
- What legal and regulatory factors in general affect this business area or process?
- What legal and regulatory factors in particular are affecting this business area or process currently?
- What are the emerging legal and regulatory risks?
- What factors have changed in the external environment?
- To what degree have these factors changed?

Focus on the internal environment. Consider questions such as

- What are the core or major activities that comprise this process or business area?
- What does the business area or process produce?
- What are the key systems and applications used by the area or process?
- What changes have occurred in staffing and/or organizational structure?
- What are the key metrics used to assess performance?
- Are these metrics still valid?
- How have these metrics changed since the prior audit?

Understand why the entity was selected for audit; for example, it is a new acquisition that's never been audited, it is high risk due to management turnover, and so on. Focus on gaining an understanding of the nature of this entity vis-à-vis the rest of the organization and its external (competitive) environment.

One challenge in risk-based process auditing is identifying the accountable process owner. From a practical perspective, most processes do not have one person responsible for their performance. Instead, processes are divided into functional segments and organized into departments or divisions. And most departments and divisions do not have clearly articulated goals; they focus on performing activities.

In the absence of clearly defined objectives, it is helpful to begin an operational audit by categorizing the process as maximizer or minimizer. It is also helpful to consider whether the process is a line, support, or control function. Additionally, each organization should have a process to create long-term strategic goals and to measure progress. Tactical objectives should correlate to and support strategic plans. These tactical plans should be measurable.

Process	Examples
Maximizer: make, acquire or generate as much as possible; there is no limit	Sales, loan origination, fund development
Minimizer: make, acquire or generate as little as possible or produce it just-in-time	Staffing recruitment, inventory, purchasing
Line: processes that directly affect the achievement of business goal	Investment management, selling, lending
Support: processes that assist line functions	Recruitment, accounting, systems development
Control: processes that mitigate risk	Reconciliation divisions, user acceptance testing divisions, quality control units

You may find it helpful to create a high-level diagram using Microsoft Visio or another flowcharting software package to depict the major activities that comprise the process under review. This high-level depiction should require no more than a single 8.5 × 11 in. sheet of paper. You can use this high-level overview diagram to segment the process into functions, which will then make it easier for you to estimate resource requirements, set the shape of the audit, and identify the risks in the process.

Before you can finalize your audit's scope, you need to gain an understanding of the process you need to audit.

WHERE TO GET INFORMATION ABOUT THE PROCESS UNDER REVIEW

You can gather information from a range of sources, including the constituent, and use this information in a thoughtful way as the audit or review is planned. Be sure to vet constituent-supplied data for accuracy and completeness.

To gain an understanding of the process, you can

- Obtain and review information from the constituent, for example, monitoring and metric reports, including key performance indicators (KPIs), policies and procedures, and so on. Consider these data within the context of competitor performance.
- Review prior internal and external audit reports. Pay attention to the scope and objectives of these reviews. Consider the nature of any findings.
- Review prior internal audit work papers, particularly those that describe control gaps and breakdowns. Review the organizational charts.
- Examine the strategic and current business plans.
- Review operational and systems flowcharts.
- Collect information concerning the entity's performance, for example, volumes, sales, staffing stability, and so on.
- Analyze financial reports.
- Read trade journals.
- Brainstorm with internal auditors who are subject-matter experts and members of the audit department's senior management team.
- Interview the area's management.
- Use the Internet and your organization's intranet and shared drives.
- Talk to the compliance, legal, risk, fraud investigation, and business-line departments in your organization.

The data you need to plan the audit will come from primary and secondary sources. Primary sources are interviews with senior local management and other informed parties. Secondary sources include prior audit reports, the business's strategic plan, the results of self-assessments, management reports, and any other published data.

Your initial questions need to target the appropriate level of information, that is, an overall understanding of the business process objectives, the entity's major functions, KPIs, major systems, critical applications, key relationships, interdependencies, how revenues are generated, key growth trends, market trends, and so on. Obtaining this information is critical if you want to understand the business, department, or function under review.

You need to balance the amount of detailed data you collect so that you have neither too little nor too much information. Too little information will make it difficult, if not impossible, to draw accurate conclusions concerning the auditable entity. Too much information provides unnecessary detail and wastes time and effort. So that you strike the right balance, consider using the following questions:

- In general terms, what is the process's purpose or mission? What are the major activities performed? What activity starts the process? What function ends the process?
- What are the business or process objectives? Are they consistent with corporate objectives?
- What departments are involved in achieving the objectives? Who is accountable for the process? Who has responsibility for the process and its activities?
- What percentage of revenue (or expense, specific transaction type, profit, etc.) does the business generate? Is this an increasing or decreasing percentage compared with prior years and future plans?
- What standards are used to evaluate the entity's performance? Are they measurable? How has the entity fared against these standards?
- What are the major systems used in the process? What are the major applications and data stores? As a result of increased use of cloud technologies, these applications and data stores are no longer under the control of the audited entity.
- In how many geographic sites is this process performed?
- How many transactions are affected by the process and its functions? What is the monetary value?
- What are the strategic plans for the entity or process? Is it a growing segment of the company? A declining segment?
- Consider the entity or process from a historical perspective: Was it always a part of the company? Was it acquired? What has been its historical performance?
- What is the external environment in which the entity operates? What factors have changed? Is it competitive? Who are the major competitors? What's its competitive placement in the marketplace? Do a few major players or many small players dominate the marketplace?

- What legal and regulatory factors in particular are affecting this business area currently, that is, what are the emerging legal and regulatory risks?
- What is the transaction and monetary volume produced by the entity or process on an annual basis? What has been the growth trend during the past several years? Increasing? Decreasing?
- How automated is the process?
- What functions are outsourced?
- How stable is the management team and organizational structure? What is the organizational structure: decentralized, centralized, matrixed? Have there been major reorganization efforts? Is the entity or process undergoing any major initiatives, for example, reengineering, quality, team management, and so on?
- Have audits of functions or processes relating to the process you are evaluating been recently conducted? Were prior reviews conducted by internal resources like compliance or Sarbanes–Oxley* teams, or were they conducted by external entities such as regulators and external auditors? What were those audit results? How much time has elapsed since the previous audits? To what extent can you rely on the work performed by these other parties?
- How important is the process to the overall results? Does it have a global risk impact? Is it significant in relation to other business processes?

Note: These questions are intentionally general in nature, enabling you to acquire a broad-based understanding of the entity and its process(es). In addition to the questions listed, you might want to ask the constituent questions about the following topics:

- The entity's culture, vision, mission, and organizational structure
- The entity's core processes and key outputs
- The entity's constituencies: customers, suppliers, strategic partners, agents or outsourced functions, and employees and the key drivers affecting each constituency
- The entity's financials: revenue generation drivers, major capital versus operating expenses
- The entity's product or service mix, including the rate and nature of new product or service development
- The entity's susceptibility to key trends in the areas of technology, globalization, government regulation, the economy, and demography
- The entity's planned and actual use of outsourcing

---

* Although the board of directors, the CEO, the executive committee, and managers at all levels have always been responsible for a company's performance, including the accuracy, completeness, and timeliness of financial reporting, recurring events in U.S. business have indicated that improvement is needed. In July 2002, Congress enacted the Sarbanes–Oxley Act. The Securities and Exchange Committee (SEC) is responsible for enforcing compliance with this law. The Sarbanes–Oxley Act was intended to increase investor confidence in a company's financial reporting process, requires increased corporate governance, clarifies the role of the external accounting firm, facilitates whistleblowing, and emphasizes risk management.

The following additional guidelines are helpful if you are analyzing the financial performance of a subsidiary or performing a financial audit:

- What are the most significant aspects and risk areas of the business?
- What are the accounts receivable statistics and turnover?
- Is there a variance from year-to-date (YTD) budget/plan to actual and from YTD actual to prior year? Why?
- What are the revenue statistics and trends? How and when is revenue recognized?
- What are the cash-flow trends?
- What are the expense trends?
- What is the nature of the accounts (assets and liabilities) and the types of balances you should expect to see?
- What are the trends in prior periods and budgeted amounts?
- Is the trial balance in balance?
- Do all detailed analyses or schedules tie to the trial balance?

When evaluating the answers to these questions, determine how the performance compares with the rest of the company or to similar functions.

When you need to identify the key systems in the area under review, it is helpful to obtain a dataflow diagram or to create one to show the data inputs, outputs, and exchanges, and whether the data are processed on a real-time, batch, or manually dependent basis. At a minimum, create a list of applications, databases, and systems. For each item, identify the area responsible for its maintenance and list the major users.

1. Consider which systems are key by thinking about the degree to which they
   a. Support the business area's key processes
   b. Receive inputs from key processes
   c. Provide outputs that are key to downstream processes
   d. Are associated with aspects of the key process that have changed
2. Communicate with key IT contacts and key business contacts to obtain their perspective or to validate existing information.
3. Gather pertinent information about the system:
   a. Branch, region, legal entity, and country location of the system
   b. The physical location (that is, the host) for the system
4. Gather information about changes since the last review or audit, including
   a. New interface
   b. Changes to the interface
   c. Changes in scripts or codes
   d. Change in system location
5. Communicate with management about the status of work related to these key systems, considering questions such as
   a. What is the most recent work done by them on the system?
   b. What type of work have they done on the system?
   c. What recent changes, if any, has the system undergone?

    d. Does the unit have any non-IT-supported applications or systems or does the unit have applications (not models) that perform calculations (including Microsoft Access databases or spreadsheets)?

    e. Can end users create and modify macros that affect the outcome of a calculation?

    f. Who is responsible for securing access to those applications and performing system administrator duties?

6. Communicate with the constituent concerning
    a. Users of the system for key processes
    b. Business and IT owners
    c. Downstream and upstream interfaces relevant to the process
    d. Age of the system and availability of upgrade versions

Sometimes the process or entity management may have initiated specific projects that will alter the process or entity's boundaries and affect the audit's scope. To gain an understanding of these project activities and plans,

1. Obtain from the process or entity management and review a list of planned and active key projects.
2. Understand each project's goals and desired outcomes. To what extent do they affect the entity's purpose and boundaries and the audit's scope?
3. Determine whether the process or entity management has performed or planned risk assessments relating to the projects or area(s) you are planning to audit. Have they conducted any control self-assessments? If so, were any issues identified? Review the results of those assessments to determine whether you may need to modify your audit approach.
4. Communicate with entity management about the status of work related to projects, considering questions such as
    a. What is the most recent work done by management on the project?
    b. What type of work has been done on the project?

If there are no scheduled or active projects, indicate this in your planning documentation.

OTHER TIPS FOR EFFICIENCY DURING THE PLANNING PHASE

1. Determine your interview objectives and identify in writing the key questions or concerns you need to address during the interview.
2. Make sure you interview the appropriate person(s).
3. Set up the meeting in advance.
4. Explain how the risk-based audit process works.
5. Inform the interviewee of the topics to be discussed and indicate that it may be beneficial to have key personnel involved in the areas to be reviewed also in attendance.
6. Establish the estimated time for the interview and confirm this with the interviewees. Be grateful for their assistance and be sincere about it.

7. Understand the business objective and how the process works before trying to assess the risks.

8. Review the prior planning materials and audit report to determine whether the audit's scope should be expanded to include processes or functions that were not reviewed in the past.

9. Determine where data reside as well as their layout and accessibility during the planning phase to identify opportunities for full population testing using computer-assisted auditing tools. Leverage the use of data analytics to understand the population(s) you will evaluate.

10. Document the results of each of your interviews, noting key control points and weaknesses and any open items that should be addressed in subsequent data-collection interviews. It is helpful if you can prepare any narratives and flowcharts soon after the interview when your memory is fresh.

11. To test your understanding of the process, draw a high-level process map. If there are major gaps in the flow, obtain more information about how the process works.

12. Make sure that maps contain only pertinent information concerning the process or control under review. Do not create "map-narratives."

13. Make sure you have walked through the process to identify the risks that can occur in each of the process's major functions.

14. Communicate with line management early and openly.

15. Make sure you have discussed and understood management's current concerns.

16. Talk with audit management to gain an understanding of executive management's concerns.

17. Be sure to include audit activities that will enable you to provide business management with conclusions that address their concerns and are supported through evidence, not just observation and inquiry.

18. Find out whether key people (auditors and constituents) will be available during the audit.

19. Meet with your audit team (assuming you are the lead) to establish expectations for project performance and to set ground rules. Find out the skills and development needs of your audit team members.

20. Find out whether any holidays will occur and what the impact will be on the audit schedule.

21. Plan to leave a cushion in your audit's hours as a contingency budget.

22. When requesting information from constituents, use the report name only if you know it. Otherwise, describe your informational objective or the type of information you want to analyze. For example, instead of requesting an aging report, ask for whatever reports they have that disclose the number and amount of unpaid invoices.

DATA-COLLECTION INTERVIEWING: THE FUNNEL APPROACH

How you ask your questions is just as important as the questions you pose during your data-collection interviews. In general, you can pose questions like the ones in this chapter via e-mail (as an initial risk assessment or internal control questionnaire), but you will acquire richer—and more useful—data if you interview your constituents in person, over the phone, or using video conferencing. Funneling is a powerful interviewing approach that enables you to get the most and best information from your interviewees.

The funnel approach is an iterative method of questioning that enables you to methodically uncover information, opinions, and perspectives. It also helps you maintain control of the interview or conversation.

To use the funnel approach, begin by planning your interview objective and list the main topics you want to discuss. For each topic, begin by asking broad, open-ended questions. Then, focus your follow-up questions to address the responses you receive, and gradually proceed to ask more and more specific closed or direct questions on that single topic. Essentially, when using the funnel approach, you are using questions to direct your interviewee from the general to the specific concerning each topic of discussion. You will use the funnel approach for each topic you need to cover (Figure 2.2).

The funnel approach consists of four phases: open-ended questions, restatements, closed-ended questions, and "what if" constructions.

Phase 1: Open-Ended Questions: During this phase, ask questions that allow the interviewee to explain or describe a particular topic in broad terms. The intent is to obtain the most information and perspectives concerning the topic. Additionally, the use of open-ended questions encourages the interviewee to talk.

The most unstructured type of question is the open-ended question. Open-ended questions allow you to get into a topic quickly and easily. Open-ended questions also allow the person to answer in any way he or she wishes. These types of questions require more in-depth answers than "yes" or "no" or a simple statement of fact.

Open-ended questions help to establish rapport and show you are willing to listen to others. They get people to think and open up, they encourage discussion, and they allow the other person to become involved in the discussion and exchange ideas. They create a more conversational tone and eliminate a sense of "interrogation."

Examples:

- "Please describe the process of ____."
- "What are the features of the XYZ product?"
- "What solutions have you tried?"
- "How are you getting along with the new person?"
- "What are your unit's activities?"
- "What prevents you from meeting the deadline?"

Phase 2: Restatement: Restatement is a method of rephrasing the speaker's message, information, or opinions in order to verify a mutual understanding of technical terms, intent, or perspective. The importance of this phase cannot be stressed

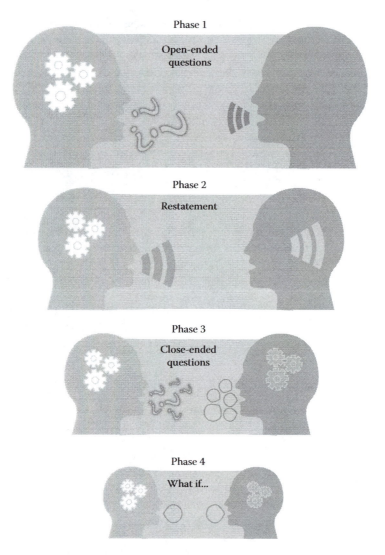

FIGURE 2.2 Funnel approach.

enough. This is your opportunity to make sure that you understand exactly what the interviewee previously stated, what actions transpired, and the interviewee's perspective and opinions regarding the topic under discussion.

You do not want to move the conversation forward until you are certain that you understand the interviewee's meaning. You can remain in the restatement phase for as long as is needed or until the interviewee agrees that your restatement is an accurate synopsis of the response.

Example:

- *Interviewee:* In the last six months, we have changed our underwriting workflow… the transaction volume is higher than it has ever been, causing us to reexamine how we do our work and who does what. The only thing that hasn't changed is our critical processing cut-offs and our underwriting standards.
- *Your restatement:* So, in the last six months, your procedures have changed because of extremely high volume. And you've had to change job responsibilities for your staff.

Phase 3: Closed-Ended Questions: During this phase, you need to ask questions that force the interviewee to respond in brief statements or commit to a "yes" or "no" position. These questions can be used to acquire more information and opinions or to verify previously received information and opinions. During this phase, phrase your questions in such a way that you keep the respondent's answers focused on the topic under discussion. The use of closed-ended questions also prevents digressions and tangents from occurring.

Closed-ended questions extract pieces of information but preclude further discussion. They are narrow questions that can often be answered with "yes" or "no" or a simple statement of fact; for example, multiple-choice questions are closed-ended.

Closed-ended questions are good for getting information or confirming agreement. They can help to keep a conversation on target or bring a discussion to a close.

They will not serve all an interviewers needs because they "close" a conversation. They do not lead anywhere as they give the interviewee the ability only to answer the question asked and not to volunteer any additional information.

Examples:

- "Would you rather meet from 10:00 to 12:00 or 12:00 to 2:00?"
- "Do you review these vouchers before or after they are processed?"
- "Are you having a problem meeting your standard?"
- "Were you able to meet your deadline last month?"
- "How many accounts can you adjust in an hour?"
- "How often do you reconcile these accounts?"

Phase 4: Conditional/"What If" Constructions: During this phase, describe hypothetical situations, cases, or examples to verify your understanding of the interviewee's message. This type of question provides you with an opportunity to explore the "gray" or ambiguous areas. Remember, the interviewee's response should be consistent with answers to prior questions. If the response is not consistent, you need to ask more specific follow-up questions to resolve the inconsistency.

Examples:

The interviewee has just described the steps taken to process a transaction, and this procedure relies almost exclusively on the availability of a computer application. Your possible conditional/"what if" construction: "So, what would happen if there was a power failure and you couldn't use the processing system?"

When preparing for your interview, remember that the person who asks the questions controls the course of the conversation. Consequently, as the interviewer, you are inherently in the power position. Using the funnel approach to address each topic you need to cover will enable you to stay organized as you collect the information you need to set the audit's scope and objectives.

PUTTING IT ALL TOGETHER TO SET THE SCOPE AND OBJECTIVES

The goal of the planning phase is to establish the audit's purpose and boundaries. Decisions made in this phase affect the nature of the work to come, including the type of testing that is performed. When thinking about the scope and objectives,

1. Consider whether or not the following issues are relevant to the business area and should be part of the audit scope:
   a. Audit issues identified during relevant prior internal audit engagements
   b. "Open" issues relevant to the business area from prior internal audit engagements
   c. Issues resulting from any other exams (e.g., regulatory exams, second defense line, control groups)
   d. Issues identified by external auditors and consultants
2. Determine the audit scope:
   a. The overall process or business area's starting and ending points or boundaries
   b. The audit period (generally the last 12 months or since the last audit, whichever is shorter)
3. Clarify at a high level what the audit is intended to accomplish by specifying what will be assessed or determined.
4. Describe areas, functions, or activities that will be excluded from the audit's scope.

Throughout the planning phase, make sure that you have open lines of communication with your manager to discuss the audit's scope and objectives, and determine whether specialized skills will be needed to complete the review. Store key information you acquire about the area in a central location, preferably in a planning checklist or audit planning scope document, so that you can easily share it with any auditor assigned to work with you. Just as importantly, establish effective communication and a decent working relationship with the constituent during planning. This will create a productive environment for the detailed assessment, testing, and reporting phases.

MANAGING YOUR TIME EFFECTIVELY WHILE PLANNING

ESTIMATING TIME REQUIREMENTS

Estimating time and resource requirements is an activity that is more art than science. The following are some techniques that will help you come up with useful and realistic estimates:

Be as specific as possible in describing the steps that need to be performed. The more specific you are, the more you will be able to imagine how long it will take to complete a particular activity.

Identify steps that require input from others, for example, data-collection activities. Check with these people to find out how long it will take them to compile the information and allow time to obtain data stored off-site.

Remember that all time spent on the audit should be charged to the audit. This will make it easier to estimate the time required to complete a similar review in the future.

Differentiate between "effort" and "elapsed" time because no one works on anything without interruptions. Effort time is the actual time a person spends working on the task or the actual "touch" time required to complete the project. For example, assume that a project activity is "to identify the trends in reason code usage." This activity may require two hours to complete once the data are obtained from various locations and systems.

When estimating the time required to complete this task, consider the calendar days required to request and receive the data, not just the actual (effort) time needed to perform the analysis.

Consider whether the area has been audited before, because the process or entity owner may not be familiar with your audit methodology. It may also take longer to gain an understanding of an area that has never been audited.

If the area has not been audited in several years, there may be process changes that need to be documented. Consider allowing time for this.

Notwithstanding historical time requirements, allow extra time if new staff is on your team for two reasons. First, it will take them longer to complete the work. Second, you will need to spend more time coaching them.

ACTIONS TO TAKE FOR BETTER ESTIMATES

Set the audit objective with care. Once the end zone is clearly defined, it is easier to set time and resource estimates. Identify the major activities. The more detailed the plan, the easier it is to develop realistic estimates.

Once the audit objectives and scope are detailed, then plan backwards to identify the major tasks that need to be performed. To do this, ask yourself, "When this project is finished, what are the last things I will have to accomplish?" Record your responses on a sheet of paper. Then ask yourself, "In order to complete these things, what was the second-to-last thing that I would have to accomplish?" Record your responses to this question. Continue this process until you have identified all the steps you will need to perform during the project. Working backwards to complete an audit, we would issue the final report; create the final report by incorporating the constituents' comments; convene constituent meetings, if needed, to resolve to problematic corrective action plans; issue the draft report; edit the report; draft the report; analyze the test results across the test plan; and so on. Once all the activities are described, it is relatively easy to come up with the time estimates for each one.

Highlight the critical dependencies. In many cases, estimates concerning the completion of some tasks rely on the prior completion of other tasks. For example, writing the audit report is a task that is dependent on completing and analyzing test results. If the test results are not analyzed, the report cannot be written, and the time required to write the audit report will be longer than estimated. Highlighting the critical dependencies enables you to more easily spot the high-risk "make it or break it" points in your project's estimates.

Document your assumptions. If your estimate includes having two auditors, make sure this is clearly expressed in your audit planning documentation and planning memo. Writing the assumptions down prevents selective memory lapses from occurring later when that resource may not be available when needed, and the time required to complete the audit is extended.

Determine the required time for each of the major activities in the planning, testing, and reporting audit phases. Since every plan needs revisions—this is a fact of life—your life will be immeasurably easier if you document the logic you used to come up with your estimates. As you make your time determinations, estimate in either half- or whole-day increments. Admittedly, the "fudge" factor allowance (technically known as float or slack) is larger if whole-day increments are used. If each of your audit projects is unique, creating a higher likelihood of unforeseen problems, using whole-day increments is a more conservative approach. If your audits are repetitive projects, using half-day increments would be more effective.

Track and use historic records to understand how much time certain activities required in the past. Estimating experience is a good thing—but only if you have a written record. Memories are too unreliable to trust. Get more disciplined about recording time usage, especially for the major audit assignments. If you know that certain audits will need to be repeated cyclically, historic information is a valuable source of estimating information. Using historic information will give you a better sense of how much time to allocate to the various audit phases, for example, planning and risk analysis, detailed control evaluation and testing, test analysis, and report issuance. Historic time usage is only valuable if the data are accurate. Get into the habit of recording data as the tasks are completed, and encourage your team to record the actual hours they work on a weekly basis.

CHAPTER WRAP-UP

You have now set the audit's scope and direction and established a positive working relationship with your team and with the area's management. You have completed data collection by asking effective questions and have made good use of your time. Whether you are responsible for planning the audit or not, you can use the Critical Linkage™ to gain an understanding of the entity under review so that you can accurately assess its risks and controls. The next chapter explains how to complete Step 2 of the Critical Linkage™: identify the inherent and external risk events.

CHAPTER SUMMARY

- Make sure you understand the critical linkage among the business objectives, major business activities, risks, and controls.
- Keep in mind that performance is relative, not absolute. An organization can get better and fall further behind at the same time.

- The link between inputs and outcomes is tenuous. Bad outcomes do not always mean that managers made mistakes, and good outcomes do not always mean they acted brilliantly.
- Demonstrate professional skepticism. This means that you are able to remain objective and question the source and veracity of information provided to you by others (beginning with the process owner whose area is being reviewed and including other auditors).
- Watch out for oversimplification and reductionism (things are rarely that simple).
- Be wary of complexities because they can obfuscate (things are rarely that complicated).
- Focus on understanding all of the process or entity's inputs and outputs.
- Determine whether any interdependencies exist between the area you are auditing and other processes or entities.
- Begin operational audits by understanding the objectives of the process under review and the major activities needed to achieve them.
- A process has clear starting and ending points and cuts across departments. Examples of processes are policy issuance, installations, making loans, taking deposits, selling nonbank products (e.g., mutual funds and annuities), new product development, and purchasing.
- Determine the required time for each of the major activities in the planning, testing, and reporting audit phases at the onset.

QUIZ

1. Audit planning
   a. Is the act of documenting all the process steps.
   b. Comprises the activities needed to acquire an understanding of the area under review, set the audit's objectives and scope, determine the resources required to complete the project, establish the time lines and deliverable due dates, and officially kick off the audit with the area's management.
   c. Is a phase only relevant to audit leads.
2. The steps in the Critical Linkage™ are
   a. Interview, document, test, corrective action.
   b. Objectives, risks, controls, corrective action.
   c. Plan, discuss, test, report.
3. To gain perspective of the area you need to audit, you must focus on
   a. The business area overall.
   b. The external environment.
   c. The internal environment.
   d. All of the above.
4. The data you need to plan an audit will only come from interviewing process owners, for example, senior local management and other informed parties.
   a. True
   b. False

5. The funnel approach involves
    a. Asking "why" until you get to the essence.
    b. Asking "who, what, when, where, why."
    c. Asking open-ended questions, making restatements, asking closed-ended questions, and using "what if" constructions.

The performance planning worksheet

My measurable, time-bound performance goal (i.e., my desired outcome)

Things I want to start doing

Things I want to stop doing

Things I want to continue doing

Observable outcomes and indicators that I've made positive change happen

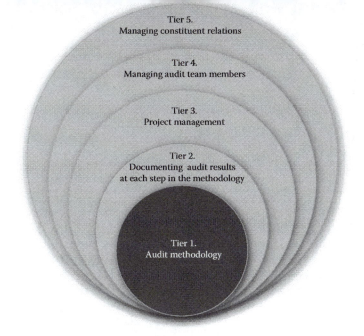

Tier 5.
Managing constituent relations

Tier 4.
Managing audit team members

Tier 3.
Project management

Tier 2.
Documenting audit results
at each step in the methodology

Tier 1.
Audit methodology

---

3 Techniques for Detailed Risk and Control Assessment

CHAPTER CONTENTS AT A GLANCE

This chapter will

- Walk through the Risk Mental Model—the second step in the Critical Linkage™
- Define the difference between inherent and residual risk
- Explore five ways to identify inherent risk events and their causes
- Explain how to use risk categories effectively
- Explain how to distinguish inherent risk from broken, missing controls, and other things
- Discuss factors to consider when assessing risk
- Understand and apply the Control Mental Model—the third step in the Critical Linkage™
- Describe criteria for evaluating the adequacy of control design

- Define control environmental elements, control activities, and monitoring in terms that your constituents will understand
- Explore the value of differentiating control environmental elements from control activities and monitoring
- Articulate the attributes of key controls
- Describe tips for conducting effective walk-throughs

Risk comes from not knowing what you're doing.

Warren Buffett

There is so much to do in the detailed risk and control assessment phase. The results of this phase will drive and shape the testing activities, which in turn will impact the contents of the audit report. But let's not get too far ahead.

During the planning phase, you should have acquired an understanding of the entity, process, or area under review and its importance or value to the rest of the organization. Based on this understanding, you set the audit's scope and determined how you wanted to segment or divide the entity, process, or area under review into components so that you could manage the audit like a project and estimate time and human resources requirements.

Having planned the audit and assuming it is an operational or process audit, the first and most important thing you need to do next is a detailed inherent risk assessment. As part of this activity, you need to identify, analyze, and assess the inherent risks that threaten the achievement of each function or segment that comprises the entity, application, or area under review. Essentially, you are implementing Step 2 of the Critical Linkage™ and leveraging your understanding of the business, functional, or project objectives and major steps that comprise them.

Now, if you are performing a Sarbanes–Oxley (SOX) review, a model audit review (MAR), or a compliance review, you will skip the risk-assessment step because the risk is already known. SOX and MAR reviews focus on how effectively the financial reporting risk is mitigated. Compliance reviews focus on the applicable regulatory and legal risks affecting the entity or process under review. Operational and process reviews focus on a broad array of risks that include financial reporting, regulatory, and legal risks, which is why the inherent risk assessment is necessary. Some of these risks can do greater harm than others, which is why we need to concentrate our efforts on the areas of greatest risk during operational and process audits.

The following are the key critical thinking activities you need to perform as part of the inherent risk assessment:

- Identify and describe the negative events (inherent risk events) that threaten the objective's achievement.
- Identify the triggers or cause(s), that is, how or why the risk event could occur.
- Define the impact, likelihood, duration, and velocity, that is, what the consequence is to the organization as a whole.

- Determine the assessment rating, that is, whether the inherent risk is high, moderate, or low.

Essentially, you'll need to answer these six core questions during risk identification and assessment:

1. What could go wrong? (Risk event)
2. Why or how could the exposure or vulnerability occur? (Cause)
3. How bad could the exposure or vulnerability be (the consequence in terms of impact or significance)?
4. How likely is it that the exposure or vulnerability could occur (the consequence in terms of frequency or likelihood)?
5. How long could the risk event last? (Duration)
6. How quickly could the risk event spread? (Velocity)

JUST WHAT IS INHERENT RISK?

Before going any further, let's define inherent risk. This is a negative outcome or event that threatens the achievement of a business objective in the absence of any action to control or modify the circumstances. It is caused by external or internal circumstances or environmental factors. Essentially, inherent risk is the naturally occurring, uncontrolled risk that exists in the entity or process. Anyone or any organization engaged in this activity would encounter the risk.

For example, let's assume that I decide to change careers. Instead of being a consultant, I want to own a hotdog and chestnut-cart concession that occupies a street corner in mid-town Manhattan. Let's say that I made this choice because I believed that the life of a street vendor was risk free. Well, you're probably chuckling because you know that street vendors face many inherent risks. Let's consider a few:

- They could get robbed.
- They could purchase tainted food products.
- They could prepare foods incorrectly, that is, undercook or overcook them.
- They could offer products the market doesn't like to eat.
- They could set the wrong price for their food—under- or overcharging.
- They could have a bad location.

Each of these risk events affects any hotdog and chestnut-cart owner. To the extent that an owner institutes measures to address and mitigate the risk, otherwise known as controls, the remaining or residual risk is reduced. However, before we can consider the effect of controls, we need to understand the nature and extent of the inherent risk. Consequently, the ability to accurately identify the inherent risk events is critical. If the inherent risk is inaccurately identified, you will not be able to accurately determine the control objectives, that is, the types of controls that are needed.

At this point in the audit, you should apply the following Risk Mental Model as a guide to the types of thinking activities you need to complete (Figure 3.1).

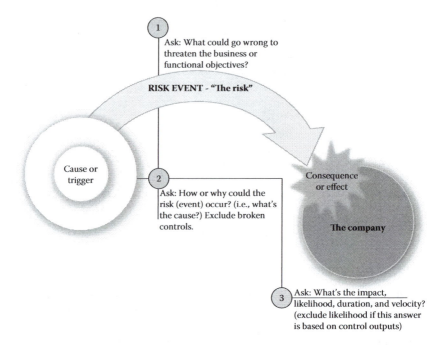

FIGURE 3.1   Risk Mental Model.

USING THE RISK MENTAL MODEL

Risk events (or simply "risks") are the bridge between risk's cause and its impact (the consequence to the organization as a whole). They identify the discrete, specific occurrences or outcomes that would negatively affect the business, process, decision, plan, product, or client (e.g., inaccurate information is provided to customers; transactions are processed incorrectly, requiring reimbursement of client accounts; transactions are recorded incorrectly, requiring restatement of client reporting). Potential risks are evaluated in terms of their impact, likelihood, duration, and velocity.

Let's focus on completing Step 1 in the Risk Mental Model. This involves answering the key question: What could go wrong to threaten the business, function, or process objective? In other words, what could impede the process or entity owner's ability to achieve his or her business goal? In order to answer this question, you need to have a firm understanding of the goal. And if you did an effective job during the audit's planning phase, you have a clear idea of what the business, entity, or process is trying to achieve. You should also understand the functions or steps that constitute the process needed to accomplish the objective.

Following are five ways to generate answers to the key question, What could go wrong to threaten the business, function, or process objective under review? Each method has advantages and disadvantages. Since our goal is to identify all of the internally and externally generated risks that could threaten the business objective's achievement, you need to use a combination of these methods. Simply using just one will hinder your ability to come up with a comprehensive list of risks.

If you struggle to come up with risks while using any of these techniques, you do not have a sufficient understanding of the business, function, or process that you are trying to audit. You will need to revisit or add to the background information that you compiled during audit planning (Table 3.1).

COMMON RISK CATEGORIES

The following are some of the more common risk categories:

Access risk: When unauthorized persons gain access to transaction processing functions or data records, enabling them to read, alter, add, or delete information resident in data files or to enter unauthorized transactions for processing.

Availability risk: When systems, or data within systems, are unavailable during the time when transactions must be processed.

Change management risk: When an organization undergoes organizational changes, including reorganizations, consolidations, acquisitions, and divestitures.

Credit risk: The risk that principal and interest will not be paid. It exists when a counterparty to a transaction owes funds to a company or is part of a financial agreement with a company. The counterparty may not pay or fulfill its part of the agreement.

Customer service risk: When an aspect of organizational conduct breaks down, adversely affecting customer satisfaction. This risk can also include an inability to modify the product or service line to suit customer needs or to provide customers with what they want.

Cyber risk: This is a group of very high impact risks once thought to be unlikely, which differ in technology, attack vectors, and means and includes all sorts of strange scenarios: organization-specific, specially designed malwares; manipulated hardware and firmware; stolen certifications usage; spies and informants; exploitation of vulnerabilities in archaic hardware; attacks on third-party service providers.

Environmental risk: When a company's operations, such as oil drilling or real-estate construction, may damage the environment.

Financial risk: When assets are subject to loss or when retained earnings may be insufficient to allow a company to grow. The loss may be tangible (e.g., theft, duplicate payments, personal use of corporate resources, fire in the corporate building) or intangible (e.g., inefficient operations, untimely investment of cash, etc.).

Foreign currency risk: When currency exchange rates fluctuate. The value of nonmonetary assets and funds will change in U.S. dollar terms with exchange rate changes. A company may experience decreases in the value of its nonmonetary assets and funds.

Fraud risk: Financial reporting—misstating company or customer financial statements.

Fraud risk: Misappropriation of assets—theft of cash, securities, or physical assets.

Fraud risk: Corruption—use of a position of trust for dishonest gain (Box 3.1).

Interest rate risk: When changes in interest rates negatively affect the value of assets (such as bonds) owned by a company.

TABLE 3.1

Five Ways to Generate Answers to "What Could Go Wrong?"

Method	Advantages	Disadvantages
Inverting the business objective	This rapidly identifies the major risks. This is arguably the most popular and quickest way to identify risks.	It masks the auditor's limited understanding of the process. More importantly, it doesn't supply sufficient detail concerning how these risks could occur, which will make it difficult to identify the key controls during Step 3 of the Critical Linkage™.
Solitary Brainstorming	This is relatively rapid and encourages out-of-the-box thinking needed to identify external risks. It is easy to do and a natural, reflective reaction when posed with the question: What could go wrong to threaten the objective?	The comprehensiveness of the risks identified is dependent on an individual's knowledge of the process under review. The more the individual is a subject-matter expert, the more the answers will be accurate and comprehensive.
Group brainstorming	This harnesses the mental diversity to generate different types of risks. If you are not working on the audit by yourself, this is a viable approach for you—one that capitalizes on the rest of the audit team's perspectives and understanding of the areas as well as the individual audit team member's work experience.	The disadvantages are the same as for solitary brainstorming. Additionally, "group think" may occur. This is a psychological phenomenon in which people strive for consensus within a group. In many cases, people will set aside their own personal beliefs or adopt the opinion of the rest of the group. Auditors who are opposed to the decisions or overriding opinion of the group as a whole frequently remain quiet, preferring to keep the peace rather than disrupt the uniformity of the group.
Process review	This is a structured approach that forces risk identification in each process step. To use this approach, you will need a high-level process map, flowchart, or data-flow diagram depicting the major steps or components of the process under review.[a] As you look at each shape on the diagram, consider its role in the process and ask yourself, what could go wrong at this point?	It focuses attention almost exclusively on internal risks. External risks may be overlooked.

TABLE 3.1 (CONTINUED)

Five Ways to Generate Answers to "What Could Go Wrong?"

Method	Advantages	Disadvantages
Risk category review	This provides a comprehensive mental jogger. You use each category as a mental launch pad to consider the applicability of the risk to the process under review. If the category is applicable, you need to describe the specific nature of the risks associated with it. Use this approach in conjunction with group or individual brainstorming.	Care must be taken to determine that the categories are clearly defined and that descriptions of how the risk could occur are explained.

[a] Very important: be sure that the diagram does not depict controls. You do not want to describe the control or detection risk at this point (you will do this during Critical Linkage Step 3).

Liquidity risk: This exists in all noncash assets owned by a company. This risk increases with the complexity of converting an asset to cash; for example, real-estate limited partnerships and fixed assets are more difficult to convert to cash than commercial paper. A company may experience delays and losses in value if it attempts to convert its assets to cash.

Market risk: When an institution's condition can be affected by adverse movements in a specific market's rates or prices, such as interest rates, foreign exchange rates, or equity prices.

Political risk: When a company commits assets to operations in foreign countries, cultural differences, acts of war, or appropriations of foreign-owned assets may affect the value of those assets, causing them to decrease in value or become worthless.

Processing risk: Risk that exists when legitimate transactions that have been input for processing or generated by the system are

- Lost
- Not completely processed or reported
- Inaccurately processed or reported
- Processed or reported in the wrong accounting period

This risk that exists in the "black box" itself reflects the possibility that the computer system will not do what it is supposed to do when it is supposed to do it.

Regulatory risk: When a company unintentionally acts outside the confines of the law, securities regulations, or banking or insurance regulations. A company may be unaware that laws exist or may have unintentionally misinterpreted the law. A company may not be in compliance with a law or regulation. If laws or regulations are imposed or changed, a company's assets may lose value or become worthless, and that company may not be able to continue selling certain products.

BOX 3.1   SOME THINGS TO CONSIDER ABOUT FRAUD

A whole book could be (and many have been) written on fraud alone. Below are some general considerations to keep in mind when evaluating risks and controls:

According to the Institute of Internal Auditors (IIA) standards:

"The internal audit activity must evaluate the potential for the occurrence of fraud and how the organization manages fraud risk." [Effective October 2012—IIA Standard 2120.A2]

"Internal auditors must have sufficient knowledge to evaluate the risk of fraud and the manner in which it is managed by the organization, but are not expected to have the expertise of a person whose primary responsibility is detecting and investigating fraud." [Effective October 2012—IIA Standard 1210.A2]

Some practical ways to put these standards into practice are

- Identifying areas of risk where theft or manipulation may be likely to occur
- Determining the adequacy of controls in accounting, financial reporting, and other areas subject to theft, fraud, or embezzlement
- Assessing the control environment and governance to make sure that an appropriate oversight process exists
- Evaluating antifraud practices and controls
- Demonstrating competency and "professional skepticism"

In every audit, the exercise of professional skepticism is paramount. In many audit failures involving fraud, inadequate professional skepticism is frequently cited as a significant reason why the material misstatement was not detected by the auditor. Professional skepticism is an attitude that includes a questioning mind and a critical assessment of audit evidence. This means that you should not assume that management is dishonest or unquestionably honest.

You should conduct the audit with a mind-set that recognizes the possibility of material misstatement due to fraud, even if no fraud has been discovered in the past and you and your fellow auditors believe that management is honest. At the same time, you should not be satisfied with less than persuasive evidence because of a belief that management is honest.

SOME WAYS TO IDENTIFY FRAUD

The following are some common ways to identify fraud:

- *Brainstorming*: Based on your knowledge of the process, generate ideas alone or with audit team members concerning how fraud could occur.

- *Following the cash*: If you understand where cash comes from and where it goes in the business or process, then you should anticipate what the income statement should show. If the picture painted by the income statement is very good, but cash is bleeding out very quickly, consider the possibility that the cash is telling the truth, while the income statement is the result of financial statement fraud. Deteriorating cash flows from operations or use of an unusually large amount of cash supposedly to fund increased inventory, receivables, property, plant, and equipment are consistent with a classic financial statement fraud, where profits and assets are overstated to conceal poor operating results.
- *Focusing on performance that bucks the trends*: Is the department performing in a manner inconsistent with that of its competitors, its industry, and general economic conditions? Perhaps it is doing very well, but only as a result of bribery or financial statement fraud. Is the performance literally too good to be true?
- *Determining whether the business rationale makes sense*: Evaluate the business decisions underlying the operating and financing decisions as reflected in the financial statements. If those decisions don't make common sense or have no clear business reason, perhaps they are motivated by hidden reasons, including fraud or illegal acts. If a department buys its leased factory building when competitors are avoiding investing in fixed assets due to poor returns, then perhaps the transaction was done to benefit the seller rather than the buyer (assuming the transaction was actually executed). If the department is reducing accounts payable during a recession, when most competitors are stretching their creditors out, perhaps those creditors are not actually being paid, but instead their invoices are being excluded from accounts payable, thus overstating profits.
- *Learning from fraud scenarios*: First, identify relevant scenarios that could potentially occur within the organization, resulting in a material impact on the financial statements. Second, for each identified scenario, describe how it would be perpetrated within the company, the individuals who could make it happen, and the financial statement accounts that would be affected. Based on the documented scenarios, identify the controls that would prevent, deter, or detect each scenario. Compare the controls in place with the controls documented and identify any gaps. Help management develop action plans to remediate significant gaps.

Reinvestment risk: When a company cannot reinvest at an equal rate of return the proceeds of a maturing investment.

Reporting risk: When management information concerning the status of projects or routine operations or financial results is reported inaccurately or in an untimely manner.

Security risk: Locations and activities may be subject to threats or destruction (physical or otherwise). A company's assets, employees, or customers may be damaged or injured.

Systems risk: Applications may be obsolete or may not do what the user requires or relies on the system to do.

Underwriting risk: When a company contractually assumes risk from other parties for certain activities in exchange for cash. A loss may occur if pricing is inadequate, expense assumptions are invalid or incorrect, changes in mortality or morbidity experience occur, or catastrophes occur.

Valuation risk: When investments are not properly valued. This risk exists when a company establishes a value for its assets or liabilities in a portfolio or financial statement. The value may be incorrect or the underlying pricing, cost, or valuation data may be incorrect.

This list of risk definitions is not exhaustive, nor is it intended to be. Additionally, some of these risks, for example, security risk, exist in all locations and activities. Other risks, for example, valuation risk and reinvestment risk, would exist only in specific functions or processes.

Early in my career, I thought having a lot of risk categories was a good thing. In fact, I amassed and shared a list of 75 risk categories with colleagues and clients. Thirty years later, my perspective has changed radically. Having too many risk categories to choose from during an audit can create confusion and bog down the risk-identification activities, adding unnecessary hours to the audit project.

The categories are intended to help you generate comprehensive and complete descriptions of the specific risks that affect the process, business, or entity that you are auditing. Essentially, this list of risk definitions is one way to start your risk identification efforts. Some of these categories will be irrelevant to the area under review. If so, just move on to the next category and think about how it affects the area under review.

Since your objective at this point in the audit is to identify all of the risks that threaten the achievement of the objective of the process or area under review, instead of using the categories to initiate your risk identification efforts, you may want to use this technique last, after you've brainstormed and used the process review approach. If you decide to do this, then you would effectively use the categories as a checklist, asking yourself, have I already considered this type of risk? If so, move on to the next category.

You may be wondering why reputational risk is not a category on the list. Damage to brand image, that is, one's reputation, is an impact. It is what happens as a consequence of the actual risk event. For example, if one is managing a call center, and the customer service representatives provide inaccurate information in a rude manner (the risk event) because they are inexperienced (the risk's cause), customers will get fed up and complain, which will damage the company's image (an impact or consequence). Depending on the nature of the negative event, an organization's reputation may be damaged.

TWO CATEGORIES TO EXCLUDE WHEN IDENTIFYING INHERENT RISK

At this point, I'd like to point out that two other categories of risk—control and detection risk—are not part of the list. This is a deliberate omission. By definition, control or detection risk is the risk that the control does not do what it is intended to do; that is, the control does not operate effectively and the nonperformance is undetected. The impact and likelihood of this risk is considered as part of the control evaluation because this risk is not an inherent one; it exists potentially because of the control structure implemented by management.

The following are some examples of control or detection risk:

Credit limits: Credit limits are no longer appropriate to minimize exposure to the organization.

New product monitoring failure: The necessary monitoring mechanisms are not adequate to monitor the new product's performance, pricing, and compliance with policies and procedures and applicable laws and regulations.

Organizational monitoring: Unsuitable corporate structure, inconsistent governance practices, unsuitable span of control, incomplete assignment of responsibility, and inadequate delegation of authority to oversee and manage organizational goals, programs, objectives, and identification of emerging risks. Inappropriate "tone at the top" management behaviors and incentives set by management for subordinates that may have a negative impact on the control environment.

Reconciliation failures: Reconciliations are not performed properly. Reconciling items are not cleared properly and in a timely manner.

While these are bona fide descriptions of things that could go wrong, they should be considered as part of the control evaluation rather than the inherent risk assessment, because they describe control breakdowns and failures of the control to perform as expected. Based on experience, I know that eliminating control or detection risk from the list of inherent risks is easier said than done.

By mistakenly adding them to your list of inherent risks, you are heading down the path of circular reasoning; the absence of a specific control type does not mean that a risk exists. Inherent risks, by definition, are descriptions of the problems; controls are the potential responses to the problems. For example, there is a risk that I may catch a cold (the problem). One of the potential solutions (controls) to this problem is that I should take vitamin C. However, it is not the only solution. Other solutions include washing my hands vigorously and avoiding contact with people who are sick. The risk is *not* that I'm not taking aspirin or that I'm not taking enough vitamin C. If we define the risk (problem) this way, then we will define the control (solution) as taking vitamin C. Also, the fact that I take aspirin does not mean that I am avoiding a cold. I may take vitamin C as a daily supplement.

Let's consider another example. Imagine that we are auditing the purchasing function. One risk event is that purchases do not meet business needs. Another inherent risk is that we pay too much for the purchases or we do not have the supplies on hand that we need to do business. An inherent risk is *not* that supervisory review is inadequate. Supervisory review is a type of control.

By describing the risk as inadequate supervisory review, the expected control would be some type of supervisory review; for example, the supervisors make sure to approve purchase requests prior to order placement. However, that would also be circular thinking. Supervisory review is just one type of control over the risk that we do not have the supplies on hand that we need to do business. Other controls that address the risk of inadequate inventory include automated controls that prevent workers from entering incorrect data such as zip and area codes (so that supplies are not misrouted) and having a list of approved vendors so that we are doing business with companies that have the capacity and inventory to satisfy our needs.

To avoid adding control and detection risks to the list of events that could nega-tively affect the business objective's achievement, take the following actions when identifying risk:

1. Understand the business objective and how the process works *before* try-ing to assess the risks. To test your understanding, consider drawing a high-level process map, one that depicts the process without controls and shows only the major functions that comprise the process. If there are major gaps in the flow, obtain more information about how the process works.
2. Make sure you have walked through the process to identify the inherent (or naturally occurring) risks that can occur in each of the process's major functions (or segments).
3. Become familiar with management's perspective on the key risks in the area. Listen carefully to determine whether they are describing inherent risks or residual risks (i.e., the amount of risk left over after controls are applied).

Let's return to the purchasing example and identify the risk events, that is, the things that could go wrong. The objective of the purchasing function is to procure and pay for supplies and services to satisfy business needs.

EXERCISE 3.1: WHICH OF THE FOLLOWING DESCRIBE RISK EVENTS?

1. Incorrect number of paper supplies is ordered.
2. Purchasing agent orders supplies late because of understaffing.
3. Incorrect payments are made to vendors.
4. Vendors do not have master services agreements in place with the company.
5. A department purchase request may be sent to junk e-mail instead of the purchasing department.
6. Items are missing in delivery shipment.
7. Processing requisitions software is down.
8. Purchasing did not send or post RFI (request for information) to potential vendors prior to placing orders.
9. Accounts payable receives an invoice for which a purchase order was not submitted.

Identifying the Risk's Cause

Once you have identified the inherent risk, consider why it could occur; that is, think about the root cause of the risk. There can be several reasons or causes. The root cause is the one that is triggering all the other reasons. Understanding whether the risk's cause is internal or external to the organization is important because it will help you determine the extent to which the risk can be prevented or detected. It will also make it easier to identify the types of controls that are necessary to prevent or detect and correct the risk should it occur.

Two examples of root causes of risk are human error and system unavailability. Let's return to the purchasing example to examine some of the causes for the risks we have identified:

EXERCISE 3.2: DETERMINING ROOT CAUSE

Risk Event	Cause
1. Incorrect number of paper supplies is ordered.	Employee hit the wrong key on the number pad.
2. Purchasing agent orders supplies late.	The department was understaffed and back logged with requests.
3. Incorrect payments are made to vendors.	The accounts payable clerk transposed the amounts.
4. Purchase request may be sent to junk e-mail instead of the purchasing department.	E-mail routing rules were set incorrectly.
5. Missing item in delivery shipment.	Purchasing accidentally deleted item from the purchase order.
6. Processing requisitions software is down.	There was a storm the night before, which knocked out power lines.

TIPS FOR ANALYZING THE INHERENT RISKS

Once you have considered the risk's cause, your work is not over. Now you need to analyze the risks—a step that is frequently overlooked. The objective of your risk analysis is to determine

- Whether the risks have common causes, that is, whether the recurring cause is automated in nature or manual.
- Whether the risks have common impacts or outcomes, that is, whether the consequences are financial, reputational, or regulatory in nature, an indicator of risk magnitude.
- The point at which the risks impact the process, that is, whether at the beginning, the middle, or the end. This is another indicator of risk magnitude.
- The nature of the risk issues that affect the process, that is, whether the underlying theme that connects the risks is failure to capture all of the transactions, accuracy, or timeliness; or a combination of all three.

The results of this analysis will help you determine whether the risks are high, moderate, or low in terms of their impact and likelihood of threatening the business objective's achievement.

UNDERSTANDING THE NATURE OF THE RISK

During this analysis, you need to make sure you understand the risk issue because the nature of the risk determines the control's objective. There are five typical operational risk issue categories:

1. Completeness, that is, the risk that we did not receive or process all of the transactions.
2. Accuracy, that is, the risk that we processed the work incorrectly.
3. Timeliness, that is, the risk that we did not process the transactions within specified time frames.
4. Legitimacy, that is, the risk that we took some action without the position power to do so.
5. Safety of assets, including people, data, and physical assets; for example, undesired changes are made to programs or processes or assets are stolen.

Depending on the business objective, one or more of these risk issues may exist at the same time. If you are auditing an operational area, the risk issues of completeness, timeliness, and accuracy tend to occur simultaneously, and all must be mitigated for the area to achieve its objectives. For example, if you were auditing call center operations where one of the goals is to provide callers with accurate and complete information in a timely and courteous manner, a risk is that callers receive inaccurate information. Another risk is that callers receive incomplete information, and another is that callers experience excessive wait times before receiving information. While these appear to be three separate risks, they are actually interrelated. From the customer's perspective, a successful service experience is one in which the information provided is complete, accurate, and timely.

Understanding the risk issue will give you an insight concerning the control objective. If you understand the control's objective, that is, the goal of the control, you will have an advantage in identifying the type of control that is best suited to address and mitigate the risk.

Understanding the risk issue also helps you determine whether the risk is pervasive or localized to one part of the process. This, in turn, will make it easier for you to determine whether the risk is high, moderate, or low in terms of its ability to threaten the achievement of the business objective.

Analyze the risks to understand the problem. Ask yourself,

- Do the risks relate to completeness, accuracy, authorization, timeliness, people management, or asset safety?
- What patterns or similarities or common causes exist among the risks?
- Which risks are pervasive?
- Which risks are triggered by external events or actions?
- Which risks are triggered by internal actions?

Make sure you secure management's agreement on the risk. During your discussions with management, focus on how these risks can occur and then discuss the types of controls that are in place.

RISK'S EFFECT: ASSESSING A RISK'S CONSEQUENCES

The ultimate effects of risks are typically financial, reputational, and regulatory. While understanding the risk's effect is valuable when you are trying to assess the risks, you need to make sure that you have done a thorough job of identifying and understanding the inherent risks first.

Once you have identified the risk events and their causes, you have enough information to assess them.

Assess the inherent risks in the process, not the residual risks. To assess the *gross* risk value, that is, the risk before the effect of controls, consider the transaction monetary value as well as the number of transactions. Also, answer the following questions:

- How bad would it be if this risk occurred; that is, what is the impact or significance on the achievement of the business objective?
- How likely is this risk to occur?
- If this risk occurred, how long would the negative condition last?
- If this risk occurred, how quickly would it spread?

PRIORITIZING RISKS: DEFINING THE TERMS

When assessing risk, it is helpful to decide the meaning of the terms *high*, *moderate*, and *low* risks before you actually begin the risk assessment. This will help you set priorities and make it easier to assess risks consistently. You may want to use the following as a guide:

- *High risk*: These are areas or initiatives that, if not properly controlled, have a high probability of creating regulatory, reputational, financial, or operational exposure of high impact to the enterprise. These can be non-routine or complex activities. New products or unique products increase the likelihood of errors. Judgment must be applied frequently. Errors are likely to have significant financial impact, require significant rework, or significantly impact the company or its customers. They could result in a significant loss of customers, brand damage, and adverse publicity. Additionally, these are areas or initiatives that are deemed "high risk" by senior management. These activities or regulatory responsibilities are highly visible, and errors are likely to have significant impact on the company or its customers and would likely result in negative and widespread publicity.
- *Moderate risk*: These are areas or initiatives that, if not properly controlled, have a *limited probability* of creating regulatory, reputational, financial, or operational exposure to the enterprise. However, that exposure could be of *high impact*. The activities are of moderate complexity. Reliance is placed on some manual processing. Errors might have financial impact, require rework, or have moderate impact on the company or its customers. These are also areas or initiatives that, if not properly controlled, have a *high probability* of creating regulatory, reputational, financial, or operational exposure but of *limited impact* to the enterprise.
- *Low risk*: These are areas or initiatives that, even if not properly controlled, create only *limited* regulatory, reputational, financial, or operational exposure to the enterprise. They are routine activities with little complexity.

Errors are unlikely to cause financial impact, would require minimal rework, and would not significantly impact the company or its customers. The activities or regulatory responsibilities are low visibility and unlikely to cause financial impact. Errors would require minimal rework and would have minimal impact on the company or its customers.

When assessing risk, consider the tendency for unaddressed moderate risks to become high risks over time. A general rule to follow is that cost and benefit considerations usually correctly lead one to focus on the higher risks and not the low risk terms. When assessing risks, do not consider the effect of controls that are in place. Focus on performing a gross risk assessment.

The following are some factors to consider when assessing the impact of the inherent risk. Essentially, the impact is the answer to the question, how bad would it be if this risk were to occur?

1. Size of assets and/or assets under management
2. Annual sales and/or revenues generated
3. Annual withdrawals anticipated
4. Volume of transactions processed
5. Size of operating budget or expenses
6. Accessibility to or ability to commit company or clients' funds
7. Importance to strategic goals and objectives
8. Severity of regulatory penalties
9. Complexity of regulatory requirements
10. Clarity of regulatory requirements
11. Level and span of customer interface
12. Experience in dealing with regulatory requirements
13. Degree of regulatory visibility
14. The enterprise's market experience/sensitivity
15. Degree of competition
16. Level and nature of technology support required
17. Availability of infrastructure required
18. Experience level of key personnel and management (also turnover in other key positions)
19. Nature and complexity of training required
20. Vulnerability to errors and fraud
21. Impact on external or internal financial reporting
22. Significant changes in operations
23. Current or planned technology changes
24. Current or planned organizational changes
25. Damage to brand image
26. Loss of customers

You do not have to think about each factor to evaluate the impact. Instead, pick three or four factors and use them to evaluate the impact of all the risks you have identified. This will enable you to come up with a consistent impact evaluation.

Assess the *likelihood* of the inherent risks by considering

Complexity: Processes that are more complex increase the likelihood of risk.

Judgment: Processes requiring judgmental decisions increase the likelihood of risk.

Manual activity: Manual intervention increases the likelihood of risk.

Interfaces: Processes with multiple departmental or system interfaces increase the likelihood of risk.

Management experience: Less management experience increases the likelihood of risk.

Organizational turnover and capacity: High turnover and low capacity increase the likelihood of risk.

Technology age: Use of legacy systems may increase the likelihood of risk if its limited features and functionality cause employees to implement work-arounds; new, cutting-edge technology may increase the likelihood of risk.

Technology features and options: The number of features and options increases the likelihood of risk.

Incorporate in your thought process

- The ways that internal and external changes in the business area may have increased or decreased risks
- The possibility that individual low-level risks may compound each other if they occur simultaneously or around the same time, resulting in moderate- or high-level risks
- The possibility that the speed with which a risk would emerge will raise the level of risk (velocity)
- The length of time that the risk could persist (duration)

Individually or in combination, significant changes to the business environment, new laws and regulations, significant loss in the customer base, major changes in the organization's processes, major changes in the organization's systems, and turnover of staff and management almost always characterize a high-risk process.

When trying to determine the frequency of a risk, do not consider the results or output of control reports. If you do, you will be evaluating residual instead of inherent risk.

When assessing the risks, consider the impact to the entity under review as well as the risk to the organization as a whole. These answers may be very different. For example, imagine you are auditing mailroom operations. The highest risk in this entity will probably not be a high enterprise-wide risk. Yet, you want the head of mailroom operations to address the risks in this area. To do this, begin by leveraging your understanding of the process under review and its importance to the enterprise overall. That will enable you to evaluate the degree to which the risk could impact the enterprise's ability to achieve its goals. Then evaluate the risk's impact on the entity's objective and determine whether it will prevent or simply hinder its achievement.

While these assessments are similar, it is important to consider both results because they will affect your next actions during the audit. If the risk's impact to the enterprise as a whole is high or moderate, you will want to determine whether controls are in place. If the risk impact to the enterprise is low, you may decide that your audit work thus far, that is, your data-collection interviewing and observation, is sufficient to support your conclusions.

Sometimes the effects or impact of risk events are mistakenly identified as the risk itself. Consider the following example concerning the vendor payment process: "Fines and penalties may be incurred when contract terms are breached and discounts may be lost if invoices are not paid." This is not the risk. This is the effect or the answer to the question, "So what? Why should this risk be mitigated?"

The risk in this example is that the company is not paying invoices or not paying them correctly or in a timely manner. Similarly, if you determine, either through observation, interviewing, or testing, that clerks are not obtaining supervisory approval on transactions above a specified monetary amount as stated in a procedural manual (all of which are controls), the inherent risk is not that the procedures are not being followed (which is another control), nor is it that approval levels are exceeded (which is another control). The inherent risk is that the transaction may not be processed correctly.

TEN KEY QUESTIONS TO ASK TO ASSESS RISK

Evaluate the answers to questions like these in terms of the process's objective, purpose, and outputs. This way, you can determine whether the risk is high, moderate, or low. As a general principle, the higher the risk, the more critical it is to have control activities in place and functioning.

If this risk occurred,

1. How bad would it be? (Severity)
2. How likely is it to occur? (Frequency)
3. Would the area still be in compliance with regulations? Does the risk deal with regulatory compliance? (Regulatory)
4. What would the impact be on your company or a department's reputation? (Image, lost market share)
5. What would be the impact on your ability to keep or obtain customers, employees, wholesalers, retailers? (Reputation, failure to keep your promises, and financial)
6. Would the accuracy of decision-making information and financial reporting be affected? (Operational, financial)
7. Would projections be impacted? (Financial, operational)
8. Would it cause a problem for another area within your company? (Operational)
9. How many transactions would be affected? (Transaction volume)
10. What is the monetary amount of the affected transactions? (Monetary volume)

I think it is easier to identify the high and low risks first by asking, what is the severity of the risk? And what is the probability that the risk will occur? The remaining risks will be moderate risks. Remember to assess the risk without considering the existence of controls.

To identify the high-level risks, begin by inverting the business objective. For example, if the business objective is "to process, record, and settle transactions in a timely manner," inverting the business objective would identify the following high-level risks to the entity:

- Transactions are not processed completely, accurately, and in a timely manner.
- Transactions are not recorded completely, accurately, and in a timely manner.
- Transactions are not settled completely, accurately, and in a timely manner.

But don't stop. Remember to consider how risks could occur at various points during the process and brainstorm ways that categories of risks could occur.

Tips for Documenting the Risks

Once you've thought about the risks that could affect the area under review, you need to document your ideas. The following are some tips for effective risk documentation:

1. Include only those risks that are relevant to the business objective under review.
2. Define the risk by describing what could go wrong, its significance, and likelihood of occurrence.
3. Make sure the risk descriptions "stand on their own" and are sufficiently clear and detailed so as to be understood by an independent reviewer.
4. Describe the negative outcome, that is, what can go wrong; for example, products or services are sold at less than cost (the process is sales); employees are over- or underpaid (the process is payroll).
5. Write a description of the causes of the risk (negative) event; for example, pricing information is incorrectly entered into the system, data are inaccurately entered into the system, or not all transactions are received or received in a timely manner. Include a description of why or how the risk could occur and understand the nature of the risk, that is, the risk issue. This will accelerate and increase efficiency in identifying and evaluating controls.
6. Do not list broken controls as risks. A "broken control" is a description of how a review, edit, or other control procedure might not work. For example, "supervisory review is inadequate" is a broken control; the risk is that the work is not performed or performed incorrectly.
7. Differentiate risk *impacts* from risk *causes*. Risk impacts answer the inevitable question, "So what?" Monetary loss, government sanctions, and bad publicity are all the impacts of risks. The impacts are the consequences or effects of a risk event on the process and organization. A risk's cause explains why or how the risk could occur and aids in understanding the

nature of the exposure or vulnerability. Simply stating the risk impact does not explain the nature of the risk and how it could happen.

Once you have documented your risks, you are ready to consider the types of controls that management has in place.

CONTROL ASSESSMENT

UNDERSTANDING CONTROLS

When I was initially introduced to auditing, I thought I would never understand controls. However, I should explain that at the time I got involved with internal auditing, operational auditing was very new. This was 1985. It was a time when most internal audit departments were primarily responsible for helping external auditors complete their annual financial statement audit. Risk assessments were not performed as they are now because the risk then was known and it was a financial reporting risk. How times have changed.

Now, the scope of internal audits entails a broad array of risk categories, and the assessment of these risks involves thought, reflection, and rationale. However, once the risks, their causes, and their consequences are understood, it is relatively easy to identify the types of controls that are required. The "shape" of the risks (which is analogous to a problem), that is, completeness, accuracy, and timeliness, determines the "shape" of the solution, that is, the control.

MANAGING RISKS USING CONTROLS

The general term *controls* encompasses control environmental elements, control activities, and monitoring. Controls provide reasonable assurance to management that what is supposed to happen actually does. Controls also minimize the likelihood of negative surprises, for example, unforeseen problems, backlogs, and so on.

Once you have identified and prioritized the risks that may prevent an area or process from achieving its business objective, you need to determine whether adequate control environmental elements, control activities, and monitoring are in place to manage or "control" these risks.

MONITORING AND CONTROL ACTIVITIES

An effective system of internal controls includes the following elements:

- *Control environmental elements*: These are factors that contribute to the creation and maintenance of an organizational climate that values ethics, quality, and sound risk management. Control environmental elements do not provide any direct assurance that what is supposed to happen actually does. For this reason, they are also called *directive controls* because they indicate the type of desired organizational outcome. Examples of control environmental elements are policies, procedural manuals, training programs, organizational charts, performance standards, and job descriptions.

- *Control activities*: These are activities intended to achieve functional or transaction objectives, for example, to make sure all transactions are processed accurately and in a timely manner. They are the procedures and tasks that are established and implemented as part of the business unit's regular activities and that mitigate inherent risks to the business.

They can be performed multiple times a day, daily, weekly, monthly, quarterly, annually, or on an as-needed basis. The proper execution of a well-designed control will help in meeting one or more of the following five objectives:

- Completeness
- Accuracy/validity
- Timeliness
- Security of assets
- Compliance with regulations

Example: At month-end, a department's control team compares cash balances in the accounting system of record with balances in the bank statement. The control team investigates differences and resolves them as needed. Management expects resolution to occur within two days of when the exception was identified.

Monitoring is an activity or set of activities for which the primary purpose is to assess and improve the performance of controls. Control activities and control environmental elements are monitored by business unit management or other groups (e.g., enterprise risk management), and corrective action is taken or modifications are made as necessary. Monitoring is accomplished through ongoing management activities, separate evaluations, or both.

Example: Management reviews a weekly report of the age of reconciling differences. In the event there are adverse trends, management takes appropriate action with the control teams and may institute additional control activities or update control environmental elements, for example, policies.

To identify controls, ask, How does management know that the business objective is achieved or that the inherent risk is either prevented or detected and corrected? Be sure to differentiate controls from ordinary process steps; controls are risk-mitigating activities.

CRITERIA FOR ASSESSING THE ADEQUACY OF CONTROL DESIGN

When assessing a system of controls, you need to consider

- The nature of the operation
- The business objective
- The risk priority
- The need for cost efficiency and time effectiveness (Figure 3.2)

Effective controls satisfy the following criteria:

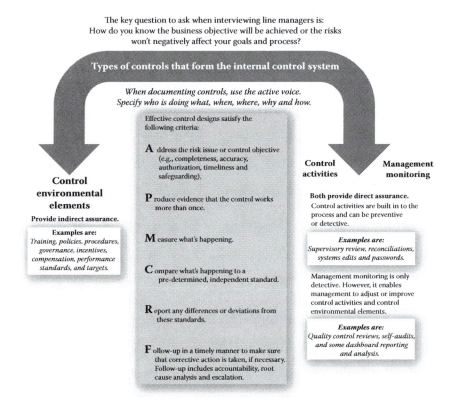

FIGURE 3.2   Control Design Mental Model.

- Address the risk issue (completeness, timeliness, accuracy, legitimacy, asset protection, and/or people management).
- Produce evidence. Business management expects conclusions that address their concerns and are supported through evidence, not just observation and inquiry.
- Measure what's happening so that management can quantify the number of times an exception has occurred. "What gets measured gets improved." When problems, errors, or defects are quantified, deficiencies can be easily traced to their root cause and corrected.
- Compare what's happening with a predetermined standard. This standard may be a law, company policy, or leading industry practice.
- Report any differences or deviations from those standards.
- Follow up in a timely manner to make sure that corrective action is taken, if necessary. Follow-up includes accountability, that is, a particular party or unit is responsible for doing it; root-cause analysis, that is, trends are identified and analyzed to determine why the defects occurred; and escalation to make management aware of the trends so that procedure and policy changes occur as needed.

TABLE 3.2

Checklist for Effective Controls

Does the control	Yes?

Answer the question, How does management know that what's supposed to happen actually is happening as expected?

Address the risk issue, for example, completeness, accuracy, timeliness, legitimacy, safety of assets, and/or people management?

Measure what's happening? The measurement is typically derived from the process objective, performance standards or guarantees, benchmarks, or laws.

Compare what's happening with a predetermined standard? The comparison typically occurs between the transaction or activity and a preexisting data store of information.

Report any differences or deviations from those standards? These reports may be manually prepared or systems generated.

Follow up to make sure that corrective action is taken, if necessary? The follow-up component is composed of accountability, root-cause analysis, and escalation.

Produce evidence that the control is working as intended?

Work in an *integrated* manner to form a system of checks and balances over a process's results?

Function to ensure that the process objective is achieved as intended?

Operate in the most cost- and time-efficient manner?

A strong control design should possess all of these characteristics. If the control's design does not meet these criteria, the control design is not adequate to mitigate the inherent risk (Table 3.2).

TYPES OF CONTROLS

Listed below are some typical controls, their definitions, and potential data-collection interviewing questions to learn more about the control design.

Controls can be considered to be *key, complementary, redundant,* or *compensating* controls, depending on the result that they produce. Understanding the difference among these types of controls is important if you want to develop effective, efficient, and reliable test plans.

Key controls are the main controls on which management relies heavily to know that the risk is being managed. Typically, key controls address more than one control objective at the same time, for example, accuracy, completeness, and timeliness. Focus your effort on identifying key controls.

Key controls can be tested by examining documented evidence using inquiry and observation, sampling, reperformance, and other testing techniques. When a test of a key control fails, you may determine that control over the process is ineffective, or you may select another control that meets the same control objective and test it. Refer to Chapter 4 for more information on testing.

Complementary controls are two or more controls that function together to achieve the same control objective. *Redundant* controls are two different controls that achieve the same control objective.

Relying on other controls to achieve the same level of comfort provided by the key control is also known as relying on *compensating* controls. A compensating control is a control procedure, not initially identified as a key control, which achieves the same objective as the key control being evaluated or tested. Compensating controls are typically ones that operate at a level of precision that is less than the key control and mitigate the risk to a lesser degree than the key control. You may decide to identify, evaluate, and test a compensating control as a substitute for a key control that is found not to be operating effectively, consistently, and continuously.

Authorization controls include approval of transactions executed in accordance with management's general or specific policies and procedures and access to assets and records in accordance with management's general or specific policies and procedures. When evaluating authorization controls, you may want to ask the control performers and their managers the following questions:

- Who performs authorization? Is it a manual process or facilitated by a computer?
- How is the authorization evidenced (e.g., signature, online initials, board minutes)?
- What triggers the need for authorization?
- Are there various levels of authorization? If so, what are they?
- What guidelines are used to determine whether authorization is appropriate (e.g., accounting policies and procedures manual, delegation of authority)? Are these followed? Is every transaction authorized?
- Is there any ability to bypass authorization, either manually or through the system? How would this be detected? Have there been occurrences of that this period?
- Would bypasses be evidenced in any way (e.g., exception/edit report)?
- Are there any management review procedures to determine that authorization is happening as intended (e.g., internal quality-control review)?

System configuration and account mapping controls include "switches" that can be set by turning them on or off to secure data against inappropriate processing based on the organization's business rules. If the switch is turned on, the checking can be customized for the particular organization to be very robust or very permissive. The more specific definition of each is as follows:

- *Configurable controls*: Specific "switches" that can be set by turning them on or off to secure data against inappropriate processing.
- *Account mapping*: Specific "switches" that can be set related to how a transaction is posted to the general ledger and then to the financial statements.

System configuration and account mapping includes standard (comes with the application or system) and customized (developed or changed by the client) controls that have been designed, based on appropriate business criteria, to secure data against inappropriate processing (by enforcing validity, completeness, accuracy) and to help ensure data integrity.

Typical examples of configurable controls are posting limits, release strategies, tolerance limits, validations and edit checks, screen layout (some fields and values are required, others are suppressed, some are prepopulated with default values, and some are "display only" values), authorization groups (as previously noted), transaction variants (a way to modify a standard transaction such as post general ledger (GL) entry, so it does something different), user parameter IDs (which populate a user's field automatically), security settings (which need to be aligned with the configuration), and configuration options (the ability to "lock" the system).

When evaluating these types of controls, keep in mind that in a "live" production environment, account mapping may be changeable by users. Mismapped accounts may not appear on the financial statements, or they may appear in an inappropriate manner such as in a suspense account or in an "opposite" category such as revenue to liability. Configurable controls can be circumvented by an end user if the control is not appropriately set up to meet the organization's needs or if user access is inappropriate. For example, using the warning message "can continue" may not be as appropriate to meet the organization's needs as "cannot continue—transaction is 'held or blocked.'"

Configurable controls can override security control features. For example, not assigning "authorization groups" to certain accounts, tables, or programs can result in ineffective security. On the other hand, a configurable control can be set up but may not be effective unless the system access supports the control as configured (e.g., a user with superuser access can just change the configured control setting).

Controls that fall into the *exception/edit report* category relate to when a report is generated by an entity to monitor something and is followed up through to resolution. In most instances, the reports are focused on exceptions/edits as defined here; however, in some instances, the content may just be ordinary information.

Exception: A violation of a set standard (e.g., customer sales exceed credit limit; three-way match does not reconcile)

Edit: A change to a master file (e.g., addition of a new employee; changes in wage rates)

When evaluating these controls, you may want to ask the control performer and their managers the following questions:

- How often are reports generated? What is the trigger for generation?
- How timely is follow-up on the report? Is follow-up documented and performed manually or via the system? If the report is system generated, does it require follow-up? Do you save reports?
- What do you look for when you follow up on the report?
- Is the exception/edit report useable? Relevant?
- Do you ever find "errors" in the exceptions reports that relate to the underlying data being wrong (e.g., the report is not giving you appropriate or accurate information)?
- Do you find errors in the report that require journal entries (e.g., the report properly highlighted something that needed to be fixed)?
- If you find errors when following up, do you also process corrective action? How is corrective action documented?
- Is there any supervisory review of corrective action? Is this documented?

Data interfaces transfer specifically defined portions of information between two computer systems using either manual or automated means or a hybrid of both, and should ensure the accuracy, completeness, and integrity of the data being transferred. The job of a data interface is to transfer the data securely, once and only once, completely, accurately, and with integrity; and to highlight any exceptions. Interfaces can be two-way (back and forth between two systems) or one-way (from one system to another) and can link new systems to old/legacy systems or old or legacy systems to new systems.

When evaluating these controls, you may want to ask the control performers and their managers the following questions:

- Are there data-integrity controls such as headers or trailers, checksums, control totals, record counts, and so on?
- What controls are in place to determine the data integrity of this information as it is being transmitted from one system to the other system?
- Are there security controls for data being transferred or converted so that no one can access them (e.g., encryption)? How does management make sure that all confidential or sensitive data are encrypted?
- Are data-management controls used, such as date or time stamps, unique file names, archiving to avoid overwriting data, and so on?
- Are there missing, duplicate, complete, or redundant data controls for inbound and outbound data?
- What controls are in place to ensure that the all of the information from one system is received completely and only once?
- What controls are used to validate or reconcile processes (e.g., online edits, batch totals, control reports, line item audit, spot checks)?
- Do you have documented procedures?

Management review is an activity in which a person other than the preparer or performer analyzes and performs oversight of the activities performed. In many instances, it will be a manager reviewing the work of a subordinate. However, it is not limited to this. It may include coworkers reviewing each other's work or one department reviewing the work performed by another department, for example, quality-control activities.

When evaluating these controls, you may want to ask the control performers or their managers the following questions:

- How often do you receive items for review?
- Do you ever initiate the review (e.g., ask to review something when it is not expected)? If so, how often does this happen, and what is usually the reaction?
- At what point in the process do you complete your review?
- What do you look for in your review? Do you have an expectation of what you'll see? From where do you build that expectation? How do you prioritize your review?
- How is your review evidenced?

- What is the timeliness with which the review is performed?
- How do you follow up on exceptions or errors if you detect them? Are exceptions or errors documented? If so, is documentation maintained?
- Do you follow-up to ensure that appropriate action has been taken on exceptions or errors? Within what time frame?

Reconciliation is a control designed to check whether two or more items, computer systems, and so on are consistent. When evaluating these controls, you may want to ask the control performers or their managers the following questions:

- How often do you prepare reconciliations?
- Are reconciliations prepared manually, by the system, or a combination of both?
- How do you know whether all base information (e.g., bank accounts and suppliers) has been captured in the reconciliation process?
- Are procedures related to preparing this reconciliation outlined in the accounting policies and procedures manual? Can I see the document?
- If the system prepares the reconciliation, do the settings or logic used reflect those in the policies and procedural manual? If not, why not?
- What source data do you use to prepare the reconciliation?
- Are reconciliations prepared by individuals who are separate from those preparing the source data (e.g., those reconciling payment information should not be the same people who are distributing cash payments)?
- What is an error on a reconciliation and how do you identify errors?
- How often do you find an error? How do you follow up? Is this evidenced in any way?
- Do you ever find "errors" in the exceptions reports that relate to the underlying data being wrong (e.g., the report is not giving you appropriate or accurate information)?
- Is management review of the reconciliation performed? How is this review evidenced?

Segregation of duties is the separation of duties and responsibilities involved in initiating transactions, authorizing transactions, processing transactions, recording transactions, disbursing funds, and maintaining custody. This type of control is intended to prevent individuals from being in a position to both perpetrate and conceal an error or irregularity. In information technology areas, segregation of duties may prevent those responsible for the development of software from releasing fraudulent applications into production without appropriate review.

When evaluating these controls, you may want to ask the control performers and their managers the following questions:

- Are procedures related to segregating duties in the accounting policies and procedures manual? May I have a copy of this manual?
- Are there system-access controls that limit an individual's ability to perform certain functions?

- What is the basis for assigning responsibilities? Is the basis documented (e.g., job description)? Is responsibility assigned automatically based on job description or can there be an override? Is there ever rotation of duties?
- Are there any management review procedures to determine that segregation of duties is adequate and happening as intended (e.g., internal audit review)?

System access is the ability that individual users or groups of users have within a computer information system processing environment as determined and defined by the access rights configured in the system. The access rights in the system should agree to the access in practice.

When evaluating these controls, you may want to ask the control performers and their managers the following questions:

- Was access established based on job responsibilities and business policy (such as clerks enter data only and supervisors enter approval data)? Does the control design consider functions that are incompatible (such as the ability to set up and approve accounts or create the ability to change and delete account information)?
- Has the system been configured to limit access of conflicting functions as part of designing the access?
- Has the design of the access limitations been configured in both the production and any other systems?

TIPS FOR WRITING ABOUT CONTROLS IN NARRATIVES

You will need to record the answers to your data-collection questions in a complete manner so that your documentation stands on its own. When documenting controls, be sure you can address the five *W*s and one *H*, that is, who's performing the control, where is the control performed, when is the control performed, why is the control performed, what is the evidence that the control was performed, and how exactly is the control performed.

1. Outline the process first. To do this, you may want to create a one-page, high-level flowchart. Then focus on describing the risk-mitigating activities.
2. Devote one paragraph to a step in the process.
3. Use the active voice.
4. Remember to describe all outcomes of a decision, that is, define the actions taken when the decision is "yes" and the steps taken when the decision is "no."
5. Use abbreviations only after the term has been defined.
6. Use bold, parentheses, or underscores to highlight control concerns.
7. Include a table of contents to make it easier to locate specific information whenever the narrative is lengthy or complex.
8. Include cross-references to other workpapers.
9. Cite the title rather than the name of the individual who is performing the action.

CHAPTER WRAP-UP

You have now covered Steps 2 and 3 of the Critical Linkage™. You learned that to perform a detailed inherent risk assessment, you need to identify, analyze, and assess the inherent risks that threaten the achievement of each function or segment that comprises the entity, application, or area under review. To do this, you need to leverage your understanding of the business, functional, or project objectives and the major steps that comprise them. You can use several techniques to identify, analyze, and assess the inherent risks to focus your audit effort on the areas of greatest exposure. Once you understand the nature of the inherent risk—that is, whether it relates to accuracy, completeness, or timeliness—you are ready to identify and evaluate the control environmental elements, control activities, and monitoring that management has put in place to address the inherent risk. You are now ready to test the controls that are designed effectively.

CHAPTER SUMMARY

- Identify and analyze the inherent risk that occurs naturally, that is, the uncontrolled risk that exists in an entity or process.
- Identify the risk events and their causes.
- Exclude broken controls when identifying inherent risk.
- Evaluate inherent risks in terms of their impact, likelihood, duration, and velocity.
- Determine what controls are in place to manage the risks.
- Evaluate the effectiveness of control design by determining whether the control activity and monitoring addresses the inherent risk and produces evidence.

QUIZ

1. A synonym for "risk event" is
   a. Risk.
   b. Cause.
   c. Impact.
2. Step 1 in identifying risk according to the Risk Mental Model is to
   a. Ask, How does management control the risk?
   b. Ask, What could go wrong to threaten the business, function, or process objective?
   c. Ask, What is the process trying to achieve?
3. Understanding the risk's root cause
   a. Is important because it will make it easier to identify the types of controls that are necessary to prevent or detect and correct the risk should it occur.
   b. Is important because it will clarify the process objectives.
   c. Is unimportant because there are many ways to control risks.

4. When assessing risks, you should consider the effect of controls that are in place.
   a. True
   b. False
5. As a general principle, the higher the risk, the more critical it is to have control activities in place and functioning.
   a. True
   b. False
6. The term *controls* encompasses
   a. Control environmental elements.
   b. Control activities.
   c. Monitoring.
   d. A and B.
   e. All of the above.

The performance planning worksheet

My measurable, time-bound performance goal (i.e., my desired outcome)

Things I want to start doing

Things I want to stop doing

Things I want to continue doing

Observable outcomes and indicators that I've made positive change happen

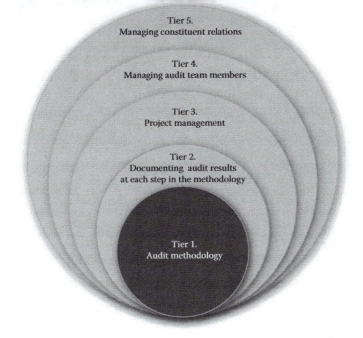

Tier 5.
Managing constituent relations

Tier 4.
Managing audit team members

Tier 3.
Project management

Tier 2.
Documenting audit results
at each step in the methodology

Tier 1.
Audit methodology

---

4 Testing and Sampling Techniques

CHAPTER CONTENTS AT A GLANCE

This chapter will cover

- The purpose of testing
- Things to consider when deciding how and what to test
- A step-by-step approach to developing and completing testing
- Options in testing techniques and the effect of this decision on test reliance
- The difference between statistical and nonstatistical samples
- Options in sample selection methods
- Factors to consider when reducing or expanding a sample size
- Tips for analyzing test results
- Ways to differentiate isolated incidents from exceptions
- Ways to identify the root cause of test exceptions
- Guidelines for documenting test work

The severest test is to learn from success.

Anon

By this point in the audit, you have acquired a firm understanding of the objectives of the area, application, or process under review as well as the major steps that comprise the functions or process. You have considered and documented the risk events that could threaten the achievement of the area's objectives. You have prioritized these risks so that you are focused on those that could cause the most damage. You understand the causes and consequences if the risk events were to occur. You have considered management's risk response, that is, whether management has decided to accept, transfer, or internally control the risk. Lastly, you have interviewed management, done some research, and observed the area in operation to identify and evaluate the design of the preventive and detective controls that management has put in place to address these risks.

TESTING TECHNIQUES

You are now ready to determine whether these well-designed controls are actually working as intended. To do this, you will test the controls to obtain evidence that they are operating as intended to manage the risks and achieve the related control objectives. Consequently, your audit efforts in this phase are concentrated on test design, sample size selection, sampling approach, and test execution. This phase should be relatively easy because of the work you have already performed. During testing, you will leverage your knowledge of the Critical Linkage™.

Essentially, testing is the examination of representative items or samples when testing every item is not possible or practical, for the purpose of arriving at a conclusion regarding the group from which the sample is selected. The results of several tests in tandem with interviews, observations, and walk-throughs will enable you to form conclusions or opinions concerning the soundness of a process or area's internal control structure. Consequently, your conclusion should relate to and is based on a result that indicates whether the control is functioning as intended.

When developing tests, you need to consider how the control is applied (i.e., automated versus manual) and whether the control is performed consistently and continuously (i.e., whether it affects all or a sample of the transactions).

Controls can be key, complementary, redundant, or compensating, depending on the result that they produce. Understanding the difference among these types of controls is important if you want to develop effective, efficient, and reliable test plans.

During this phase, you will concentrate your efforts on testing the key controls that address the high inherent risk. *Key* controls are the *main* controls on which management relies heavily to know that the risk is being managed. Typically, key controls address more than one financial statement assertion at the same time, for example, recording, completeness, cut-off, timeliness, legitimacy, validity, or measurement. Key controls can be tested by inquiry (to determine design effectiveness only) and by examining documented evidence using observation, sampling, reperformance, and other testing techniques.

When a test of a key control fails, document the control deficiency as a reportable condition. You may determine that control over the process is ineffective, or you may examine another control that meets the same control objective. Either way, you will document the result of the failed test because it is evidence that the control did not operate effectively.

If you decide to test additional controls, you might select complementary or redundant controls. Complementary controls are two or more controls that function together to achieve the same control objective as the key control. Redundant controls are ones that achieve the same control objective as the key control.

Relying on other controls to achieve a similar level of comfort to that provided by key controls is also known as relying on compensating controls. A compensating control is a control procedure, not initially identified as a key control, which achieves the same objective as the key control being evaluated or tested. Compensating controls are typically ones that operate at a level of precision that is less than the key control and mitigate the risk to a lesser degree than the key control. You may decide to identify, evaluate, and test a compensating control as a substitute for a key control that is found not to be operating effectively, consistently, and continuously. Or, you may decide no further testing is needed to reach a conclusion based on the other evidence you have collected during interviews, observations, walk-throughs and other control test results.

TYPES OF TESTING

There are two types of tests: tests of controls and substantive tests.

Control tests (also known as attribute or compliance testing) examine particular attributes or characteristics of transactions to determine if the controls are in place and operating effectively. For example, if you were testing a reconciliation control, an attribute might be a set of initials on a reconciliation. This signature is an attribute or the evidence that represents that the reconciliation was performed and reviewed. Control tests evaluate how well the internal control system prevents or detects errors. This type of testing satisfies Sarbanes–Oxley Section 404 testing requirements. While this is the most prevalent type of testing in internal audit departments, substantive testing is also used.

Substantive tests focus on the transactional details and are intended to detect and estimate misstatement in a financial statement, account balance, ledger item, or transaction. This type of testing is useful when you are auditing a money-movement process (e.g., any disbursement or revenue generation process) and the internal controls are nonexistent or not operating as intended. Substantive test results will help you size the problem created by the control deficiency.

When deciding how to conduct testing, you should consider the Critical Linkage™ and base the type or degree of testing on the audit objective and the amount of risk associated with the control that is being tested. The amount of effort you expend during testing should be commensurate with the inherent risk and control design assessment. Controls determined to be ineffective do not require testing unless you need to gauge the magnitude of the issue or quantify the impact of a known weakness. And, if you do, substantive test results will provide the answer.

Tests will determine, verify, or validate the adequacy of the control operation for those controls deemed to be well designed. You essentially have three choices when determining your test approach. You can test the entire population (known as full population testing), target your testing to focus on high-risk areas the control is intended to address (known as targeted selection), or use sampling. Your decision will be based on the characteristics and size of the population you intend to test and the degree to which audit software can be used, as well as the amount of risk the controls are intended to mitigate.

Regardless of whether control or substantive testing is performed, it is important to collect sufficient, competent, relevant, and useful information.

Sufficient information is factual, adequate, and convincing, so that a prudent person would reach the same conclusion as the person who performed the test.

Competent information is reliable and best attained through the use of appropriate monitoring and test methods.

Relevant information supports risk management issues and action plans and is consistent with the business objectives, inherent risks, and control objectives.

Useful information helps you to draw conclusions concerning the control's ability to operate effectively to mitigate the inherent risk.

Regardless of whether you plan to conduct controls or substantive testing, it is important to keep in mind the following characteristics of effective tests:

- The test must be possible, practical, and supportable; that is, test data should be readily available.
- The test results should provide meaningful evidence that specific control activities are or are not functioning as intended. You need to identify applicable test objectives before determining the test's scope, approach, and sample size. Your tests should be designed to conclude whether the control is operating effectively throughout the audit period.
- The test should represent adequate coverage for a given population to support plausible conclusions. Some potential considerations when testing are the data's location, the number of locations at which the control is performed, the number of teams performing the control, the number and type of products affected by the control, the testing period, the population, and the sample method.
- The test should be designed and documented so that it can be replicated or reperformed by someone else and that person will arrive at the same result; that is, another individual can perform the same test based on how it was originally performed.

You only need to perform tests on well-designed key control activities that satisfy applicable control objectives, for example, completeness, accuracy, and timeliness. And, instead of developing three separate tests—one for each control objective—your tests should address multiple assertions or control objectives whenever possible, because this will increase audit efficiency. It will also make it easier for you to conclude on controls' operating effectiveness relative to the inherent risks they mitigate.

A STEP-BY-STEP APPROACH TO DEVELOPING CONTROL TESTS FOR OPERATING EFFECTIVENESS

Step 1: Determine the test objective by keeping in mind that

- These tests need to identify the presence of exceptions or errors.
- These test objectives need to be answered "pass or fail" or "yes or no."

Make sure that the test objective examines

- How the control has been applied
- How consistently the control has been applied throughout the period (which should generally be the same as the audit period)
- By whom the control has been applied; that is, whether the control performer possesses the right skill, experience or competency

Step 2: Select the controls to be tested. Reflect on the Critical Linkage™, think about the risk issues and control objectives, and base your selection on the work you performed to identify and evaluate the key controls for each process. At a minimum, select the well-designed controls that address the high and moderate inherent risks that exist in the process. Make sure that there are sufficient controls for all of the pertinent objectives.

Step 3: Review walk-through documents, flowcharts, or process maps that depict how the control is supposed to work to make sure that you understand the control's purpose and design. Consider each of the following factors, basing your judgment on the Critical Linkage™ and the information collected during planning:

- Inherent risk of error
- Nature and impact of errors the control is designed to prevent or detect
- Frequency and complexity of the control
- Changes in the volume and/or nature of transactions
- Changes in personnel involved in the control
- Competence of personnel involved in the control
- Effectiveness of governance and/or monitoring
- History of errors and/or control breakdowns
- Degree of reliance on other controls (i.e., the degree to which this control relies on other controls to operate effectively)
- Manual versus automated operation
- Changes to applications supporting the controls
- Characteristics of the population
- Reliability of underlying data
- Degree of assurance provided by the sample and other tests of controls

Step 4: Determine the type and nature of the control being tested. The type of control will drive the type of test. The following are examples of typical control types:

- Reconciliations
- Management reviews
- Variance analyses (comparisons with standards)
- Monitoring
- Authorization
- Analytics

The *nature* of a key control is determined in several ways:

- Is it *manual* or *systematic*? If the key control is systematic, consider whether it is an application or a general IT control.
- Is it *preventive* or *detective*? A preventive control forestalls or prevents an error from occurring, while a detective control identifies and corrects an error that has occurred.
- Is it *monitoring* or *transactional* in nature? When differentiating monitoring from transactional controls, specifically those that are detective in nature, pay attention to the way management uses the control's output. A monitoring control tests the control objective for accuracy, timeliness, validity, and completeness and is characterized by the comparison of information or assets to policies, or it obtains verification that a control has occurred. In contrast to transactional controls, the results of monitoring are used by management to evaluate the internal control system: Should policies be modified? Should the control activities be tweaked or changed to better respond to the inherent risk? The output of transactional controls is used to correct errors or processing defects. A transactional control (also known as a control activity) occurs during routine data handling to validate the transaction's legitimacy and make sure accurate processing and recording occur.

Step 5: Identify the attribute(s) to be tested and consider whether these attributes will satisfy the test objective. If there's no evidence that the control is performed, that is, there is no attribute, reconsider why you think the activity is a control and how its design is risk mitigating. When no control exists and the risk is high, consider performing substantive testing instead.

Step 6: Determine the nature of the testing (i.e., the test method), usually by combining techniques:

- Inquiry
- Inspection
- Observation
- Walk-through
- Reperformance
- Substantive

Step 7: Use judgment to determine the extent of the testing, since different types of tests produce different levels of evidence. For example, direct evidence—that is, that which you obtained through actual observation, physical inspection, or reperformance—provides a higher level of evidence than information described to you during a conversation or interview, that is, indirect evidence. The selection of a particular type of test depends on the test's objectives. A single test, if designed efficiently, can be used to obtain evidence concerning multiple aspects of a process's internal control system. Be sure to design test procedures that are sufficiently extensive to provide reasonable assurance that the control operated effectively throughout the period.

When designing test steps, choose your test with care, because different types of tests produce different levels of reliability. Reperformance testing will provide the most reliable evidence that the control is operating effectively (Box 4.1).

Step 8: Identify the population for which the test will be performed:

- Consider whether it is practical or feasible to test the entire population if it is small or if the data are available electronically and you can use audit software.
- Verify the accuracy and completeness of the source of the population data.
- Develop a sound rationale for the population selected, including such factors as the degree of assurance provided by the sample and other tests of controls, the assessment of the risk of control failure, and the reliability of data.
- Determine the range (boundaries) of the population.
- Determine the size of the population.
- Identify the location (source) of the population. Is the data accessible electronically or only in hard copy?

BOX 4.1   TYPES OF TESTING

Here are three common types of tests: control, substantive, and reperformance.

Control tests (also known as attribute testing): Control tests examine particular attributes of transactions to determine whether the controls are in place and operating effectively. Control tests focus on how well the internal control system prevents or detects errors. This type of testing satisfies Sarbanes–Oxley Section 404 testing requirements.

Substantive tests: Substantive tests substantiate the fairness of a financial statement, account balance, or ledger item. These tests are designed to detect errors in financial statements or account balances. This type of testing is used during year-end financial statement audits.

Reperformance tests: Reperformance tests confirm that the transactions being reviewed resulted in the correct outcome (i.e., the judgments and conclusions reached by the business under review were appropriate).

- Consider the value of stratifying the population to focus on higher dollar amounts or riskier transactions, that is, targeted selection.

For an automated control, determine if the walk-through can serve as the control test itself. When testing automated controls, be sure to test every possible outcome produced by the automated control. For example, if the automated control is supposed to approve transactions under one set of circumstances or facts, deny a transaction under another set of circumstances or facts, and pend a transaction under a third set of circumstances or facts, be sure to test each set of circumstances or facts to make sure that the control does what it is supposed to do.

Step 9: Develop test steps. The scope and extent of testing should reflect the business objectives that need to be accomplished. When formulating test procedures, address or consider the following questions:

- What is the control objective?
- What will the test prove (i.e., what is its purpose)?
- Is there a reasonable mix of preventive, detective, transaction, and monitoring controls that will be tested?
- What information, documents, and reports are available for testing?
- When and by whom are the documents and reports produced?
- What information appears on the documents and reports?
- Will the result of the test fulfill the objective? How reliable will the results be?
- The type or degree of testing is based on the audit objective and the amount of risk associated with the control that is being tested. Tests will determine, verify, or validate the adequacy of controls. Regardless of the type of test performed, the auditor must collect sufficient, competent, relevant, and useful information.
- When specifying test steps, choose your testing activities with care. Different test activities produce different levels of reliance.

Use of standard terms will aid you in creating specific test steps. These standard terms indicate the degree of testing and the level of evidence that will result. Examples of standard terms are

1. *Analyze*: To break into significant component parts to determine the nature of something.
2. *Check*: To compare or recalculate in order to establish accuracy or reasonableness.
3. *Confirm*: To prove to be true or accurate, usually by written inquiry.
4. *Evaluate*: To reach a conclusion as to worth, effectiveness, or usefulness.
5. *Examine*: To look at or research closely and carefully for the purpose of arriving at accurate, proper, and appropriate opinions.
6. *Inspect*: To physically examine items, for example, assets.
7. *Interview*: This means simple testing through interviews, discussion, and other methods that provide indirect evidence that the activity under review is being performed according to the established standards.

8. *Investigate*: To ascertain facts about suspected or alleged conditions.
9. *Observe*: This means simple testing by viewing the performance of the activity.
10. *Review*: To study critically.
11. *Scan*: To look over rapidly for the purpose of testing general conformity to pattern, noting apparent irregularities, unusual items, or other circumstances appearing to require further study.
12. *Trace*: To obtain source documentation and track the transaction to the system of record (e.g., general ledger, Bocomp).
13. *Verify*: This means more in-depth testing, including the examination of documents or other physical evidence related to the activity. Conclusions are drawn based on the examination and analysis of records produced by the process under review.
14. *Validate*: This means corroboration of the correct performance of an activity by performing recalculations or reestablishing proofs.
15. *Vouch*: To prove or substantiate; to verify by examining documentary evidence; to trace from the system of record (e.g., general ledger) back to the original source document.
16. *Walk-through*: To simulate or recreate the steps used to process a transaction by literally "walking through" each process step with the people who perform the work for a specific transaction selected by the auditor. Copies of pertinent documentation that support the control points are collected along the way and saved with certification documentation. Walk-throughs should not be confused with "talk throughs." Talk throughs, a form of inquiry, are conversations with management to discuss how a control works.

Step 10: Select a representative sample in the event that you cannot test the entire population (sampling is discussed further in this chapter). The methods of selecting the sample will vary depending on the nature of the control and your test objective but, regardless of the method used, the sample should be representative of the population. If the population consists of six different transaction types, all six types should be reflected in the sample. Similarly, if the control is performed by three different managers, the sample should reflect transactions handled by all three.

If you are testing several attributes at once, that is, conducting a multiattribute test, each sample selection should be representative of and used for more than one attribute.

In the event that you cannot locate an item selected for testing—for example, it is missing from the file—consider this to be an exception and record it in your workpapers.

Step 11: Execute the test steps by comparing the sample that you selected (or the entire population if it is small or you are using audit software) with the description of the key control in the narrative or control map. Based on the result of this comparison, you need to determine whether there is evidence that each attribute is satisfied. If an attribute is not met, it is an exception and you need to find out why it occurred.

Step 12: Document the test results. These results, including any exceptions, need to be documented at a detailed level in the workpapers by maintaining an electronic file of the source documents used in the test. Your test documentation should also include the sampling information. Your objective is to document the test so that someone else could reperform the test and arrive at the same conclusion (without having to ask you for an explanation or additional information). Test documentation requirements are described later in this chapter.

Step 13: Evaluate the results of the test. You will need to consider all of the exceptions and determine whether an exception is an isolated incident or whether it is indicative of a pervasive control gap or breakdown. This determination will help you make an assessment as to whether the tested control was effective. Based on your analysis, you may conclude that the control is or is not operating effectively or is operating effectively only under certain conditions. You need to record all exceptions in the workpapers.

As part of this evaluation, you need to determine the root cause of any control breakdowns or defective transactions. Understanding the root cause is necessary if you want to be able to determine whether management's corrective action plans will resolve the control deficiency.

AUDIT SAMPLING TECHNIQUES

By definition, sampling is the application of an audit procedure to less than 100% of the population. It enables you to evaluate audit evidence about the characteristics of the items selected for testing to assist in forming a conclusion about the population. When sampling is used to form a conclusion about the population, a risk exists that the sample results are not representative of the entire population. You can reduce this sampling risk by clearly defining the test procedures, selecting the most useful and relevant type of sampling, and accurately applying the sampling methodology.

Basically, there are two types of sampling: statistical and nonstatistical. Statistical sampling is mathematically derived and will enable you to quantify the sampling risk (i.e., error rates and confidence levels) and extrapolate the test result to the entire population affected by the test. Nonstatistical sampling describes any testing method that does not measure sampling risk regardless of how rigorously and objectively the sample items are selected. Whenever you use judgment to select a sample, for example, judgmental, haphazard, or targeted selection, you are using nonstatistical sampling.

Internal auditors use nonstatistical sampling because their efforts focus on testing controls. Frequently, more than one key control addresses the risk, so that the potential effect of sampling risk is offset by testing multiple different controls that address the same process risk. The results of control attribute tests do not provide evidence concerning the accuracy of account balances or the financial statements. They provide evidence concerning the effectiveness of the internal control system.

Once you understand the value of statistical and nonstatistical sampling relative to your audit objective, you can focus on how you want to select the individual items in your sample. You have the following options:

Judgmental sampling is used to deliberately include or exclude items from a sample. For example, this can be used when the population can be segmented or stratified (e.g., small versus large transactions, equities versus derivatives, foreign versus domestic transactions). When you use judgmental selection, you select only those sampling units that meet specified criteria. You consciously influence the selection to achieve certain objectives; normally, the sample is expected to include the kinds of items that are most likely to contain errors. For example, let's assume that you are going to test a manual control. If you have a theory that this control will break down or fail to operate under certain circumstances, for example, peak volumes or heavy vacation periods, you would use judgmental sampling to choose the items you will test. Your goal is not to extrapolate the results of this test to the entire population; you are trying to prove whether the control will break down under specific circumstances. Similarly, if you wanted to focus on certain types of transactions or certain dollar amounts, you might decide to stratify or tier the population and only examine items within a particular strata. This is another form of nonstatistical sampling. Judgmental sampling should be applied when a control applies to transactions of significantly varied risks or when there are discernible circumstances that could hinder operating effectiveness.

Random selection is a technique in which each element of the population has an equal chance of being selected. Random sampling is repeatable and reduces sampling bias because the items are selected for sampling based on the use of a software-generated random number.

You should use random sampling when controls are expected to be consistently applied to all items in the population and the inherent risk of each item in the population is similar, that is, you don't perceive certain transactions to be riskier than others.

Haphazard selection is a technique that is used to select a sample that is expected to be representative of the population. The sample is selected without deliberately deciding to include or exclude certain items. The disadvantages of haphazard sampling are lack of repeatability and inadvertent bias in sample item selection.

Haphazard sampling is applied as an alternative to random sampling. It is used when the population cannot be easily identified or when you do not have access to a tool that can generate a random number to initiate sample selection. When using haphazard sampling, you choose items as though you were blindfolded and pick in an unbiased way. That's the definition, but in practice it is hard to select items without a bias. The reality is that you may be drawn to higher dollar amounts or certain days of the week or certain locations or types of transactions. To counter the bias tendency, make sure that you have a sufficient number of each type of transaction before you begin your testing. Alternatively, reconsider whether you should use judgmental sampling to select the items.

Sample Sizes: How Many Are Enough? If it is not possible for you to examine the entire population, you need to pick a sample of transactions. So, once you have decided on the method you want to use to select the items you want to examine, you need to think about how many you need to sample. This decision should be based on the frequency and nature of the control. If the control is automated, a test of one for each expected outcome is sufficient. If the control is performed several times a day and the population is large, that is, more than 1000 items, once you pick 30–40 items you should have captured a sample that is representative of the population and provides enough coverage so that you can draw a meaningful conclusion. Selecting more than 40 items from a large population will not yield added variability to the nonstatistical sample or increase this sample's representation of the entire population.

Documentation Requirements: Whatever you decide to test and however you decide to test it, you need to document the actions you take. Your workpapers document the audit work performed. They are the record of information obtained and analyses conducted during the course of an audit. Your workpapers provide the support for the audit report that is released at the end of an audit, the basis for the supervisory review of the work performed, and a means for external auditors to evaluate internal audit work. Your documentation must be able to stand on its own and allow a reviewer to understand the workpapers without talking to you.

Good testing documentation includes

- Summaries of testing performed, that is, the test workpaper, including the results, conclusions, and an exception summary.
- Detailed testing information (documented in the test workpaper). Include enough information for another auditor to be able to identify the samples selected, the attributes tested, and each test's specific results.

Comprehensive test documentation functions as an "audit trail," allowing reviewers to evaluate the action you have taken as well as the usefulness of the test and its results. Your test workpapers are used to document specifics about tests conducted and should include

- Statement of the test's objectives and purpose.
- Scope period under review (which should be the same as the audit period unless the control was in place for less time).
- Population's size and source of this information.
- Sampling methodology chosen.
- Population characteristics.
- Population location.
- Sample size.
- The criteria or attributes used to select the sample, including decisions to stratify the sample.
- Specific test procedures conducted—written in a step-by-step sequential manner to facilitate reperformance. If documenting audit software information, include specific references to audit software that was run or developed, the names and titles of individuals interviewed, and the results of software runs.
- Results of testing (the specifics concerning detailed testing conducted will likely be recorded on spreadsheets).
- Test results relating to test steps that do not require sample testing, such as detailed walk-throughs and analytical reviews, should also be documented clearly, concisely, and completely. Detailed walk-throughs should focus on key controls. Results of analytical reviews should be logically supported by the underlying source documentation.

- Attribute legend. This is a definition of each attribute (what you are testing) and a description of the characteristics that enable you to assess completeness, accuracy, timeliness, and so on.
- Tick marks and tick-mark legends, if appropriate. Tick marks are symbols that auditors use to evidence that a test has been performed and to reference specific notations or comments that the auditor has regarding the attributes being tested. Tick marks used by an auditor should always be explained in a tick-mark legend, for example, "No Exception Noted," or "Exception Noted" followed by a unique number associated with that test exception: "Exception Noted (#1)," "Exception Noted (#2)," and so on. To make it easier for others to review your work, you may want to use red bold text when describing the exceptions. A single tick mark should be used when the same type of observation applies to multiple items in the sample.
- If you determine that an attribute is not applicable and mark it "N/A," you should always include an explanation (and be prepared to discuss this with the reviewer). When N/A items could not be tested for specific reasons (e.g., they are a rare occurrence, it is not cost-effective to identify population), explaining the N/A is generally sufficient. If you need to (and can) expand a sample to execute testing, you need to explain your rationale, provide a description of the expanded sample items, and document the testing you performed. Additionally, you should amend the scope section of your testing workpaper to identify the replacement sample.
- Detailed explanations, which should be written in a concise, active voice.
- Overall conclusion as to the adequacy of controls.

You should maintain supporting documentation for the tests you performed. This documentation could be copies of journal entries, screen prints, reconciliations, and all other information used to perform the test. All of these documents should be included with your test workpaper.

In some cases, it will not be possible to include electronic copies of supporting documentation. In these cases, hard-copy files should be maintained off-line. These hard-copy files should be adequately cross-referenced to the file so that any auditor on your team can follow the work that was performed. Cross-referencing allows others to easily review the workpapers and also keeps workpapers neat and organized.

Cross-referencing shows the relationship between various online documents in workpaper databases as well as to off-line hard-copy workpaper files. It allows the project leader to easily review the workpapers and also keeps workpapers neat and organized. Keep cross-referencing simple.

The following documents, if used in testing, should be cross-referenced:

- Control assessment workpaper, that is, the narrative or flowchart, cross-referenced to the risk and control matrix
- Audit procedures cross-referenced to the corresponding test workpaper

- Audit software documents or printouts cross-referenced to corresponding test workpaper(s)
- Test workpaper cross-referenced to the corresponding supporting documents (electronic or hard-copy)

If you use a specific document, like an emerging or developing issues grid, as the summary for all test exceptions, each time you note an exception in the test results, you need to create a numerical cross-reference to this test summary document. On the test summary document, similar exceptions, for example, those with common causes or symptoms, should be grouped together.

Be sure to document testing information (including results and conclusions) as soon as possible after tests are completed to allow the project leader to review workpapers in a timely manner and provide any comments or questions that he or she might have (in the form of "review notes"). Respond to review notes as soon as possible because you might need to do additional testing, ask your constituent additional questions, or provide additional perspective in your workpapers. All review notes should be retained and returned to the project leader with your response, and all review notes should be cleared before the report is released.

Evaluating Test Results: Once you have documented your test results, you need to interpret their meaning. The following are 10 tips for evaluating test results.

1. Review the test methodology to be sure that you executed it accurately.
2. Do not extrapolate to an entire population the results of skewed or stratified samples.
3. Consider the meaning of the test results to answer the question: So what's the significance? Determine the significance of the errors by considering the effect that the errors will have on the area's performance. To make this determination, refer to the inherent risks and their consequences or impacts.
4. Evaluate the results of the test against standards, leading practices, and policies. Consider the nature of the gap. How big is the gap between what should be happening and what is actually occurring? This will enable you to evaluate the residual risk.
5. Review the results for relevance to the control purpose. Is there a pattern to the attributes that were not met?
6. Consider whether the deviation is the result of a random error or whether the errors are pervasive.
7. Review the exceptions to identify root causes; that is, why did these exceptions occur?
8. Look for patterns or trends among the exceptions; for example, did the same person or team perform all the exceptions?
9. In cases where there are several controls addressing a common risk, consider the results of other tests to determine whether there are similar results.
10. Periodically, look for trends among the results of several tests to obtain perspective concerning the area under review.

When analyzing test results, you need to differentiate isolated incidents from exceptions. An isolated incident is an aberration or deviation that is infrequent, a one-off, or insignificant. An exception is an aberration or deviation that is systemic, widespread, and/or material.

When trying to differentiate isolated incidents from exceptions, a common reaction is to increase the sample size. This is not the most efficient practice. Instead, carefully review the test methodology to ensure that it is constructed in a useful, meaningful way. Then, consider the results of other tests to detect trends among results. Also, consider the control environment over the process under review. To determine significance, ask yourself, What is the risk to the business if this control isn't functioning, and what is the impact or effect of the test results on the process?

Understanding the cause of the condition, that is, your test result or what you found, will help you determine whether your result is a one-time aberration or something that indicates a systemic problem. Essentially, you are trying to identify the factors, activities, or events that caused the condition. Insofar as a condition may have been triggered by several causes, you want to identify the root cause or the one that started the whole cause-and-effect chain reaction. To do this, you need to differentiate symptoms of the problem—that is, the conditions or situations that exist that were identified by your test results—from the root or base event—that is, the reasons why the test results, the conditions, or the situations exist.

The purpose of this root-cause analysis is to isolate and resolve the circumstances or factors that trigger a problem, control breakdown, or risk issue. Failure to address the root cause of a problem or risk results in wasted time and resources. The ability to accurately identify the root cause of the risks is critical. If the root cause is inaccurately identified, you will not be able to recommend appropriate and cost-effective control activities and monitors.

The following are four ways to identify the cause of exceptions:

1. *Trend analysis*: This approach enables you to identify the patterns or similarities among incidents or transactions. However, simply using this method is insufficient to determine the root cause.
2. *Ask "why" five times*: This type of questioning enables you to drill into the root cause and get beyond the description of the symptom.
3. To differentiate causes from symptoms and to identify root causes, ask, *If you fixed the situation, would the problem be fixed completely and not recur in the future?* If the answer is "yes," you have identified the root cause.
4. *Focus on the real people, places, and facts*: This means talking directly to the people involved, visiting the actual work site, and collecting factual data, not opinions or assumptions concerning the risk, problem, or situation. Involve line management in discussions concerning the cause. They are frequently in the best position to explain why something is or isn't happening in this area. You need to understand the cause in order

to recommend the most effective corrective action plans. Causes, when reported, must be supported with sufficient evidential matter in the workpapers.

CHAPTER WRAP-UP

Some would argue that testing is the core of auditing. It is the bridge between concept and reality. It is how you are able to demonstrate to management and the board that controls are or are not working as intended. This chapter has covered how to logically approach and plan tests, how to efficiently select sample sizes, and how to document and interpret results. When test results are well written, the next step, writing the audit report, is relatively easy.

CHAPTER SUMMARY

- The testing phase includes test design, sample size selection, sampling approach, test execution, and test result evaluation.
- There are two types of tests: tests of control and substantive tests.
- You need to understand the root cause of all test exceptions so that management's corrective action plans resolve the control deficiency.
- Sampling is the application of the audit procedure to less than 100% of the population.
- When you cannot test the full population, you can use judgmental, random, or haphazard sample selection.
- Your documentation, which functions as an audit trail, must allow a reviewer to understand the actions you took and to reach the same conclusion without talking to you.

QUIZ

1. When determining a testing approach, you can
   a. Test the entire population.
   b. Focus on high-risk areas.
   c. Use sampling.
   d. All of the above.
2. The first step when developing control tests for operating effectiveness is
   a. Selecting the controls to be tested.
   b. Gaining an understanding of the control design.
   c. Determining test objective.
   d. None of the above.
3. Sampling is a mathematically derived testing procedure that will enable you to quantify the sampling risk.
   a. True
   b. False

4. Substantive testing is never used.
   a. True
   b. False
5. Regardless of how many times the control is performed and the population size, it is best to pick 30–40 items or the most possible.
   a. True
   b. False
6. An isolated incident is
   a. The same as an exception.
   b. An aberration that is infrequent, a one-off, or insignificant.
   c. An aberration or deviation that is systemic, widespread, or material.

The performance planning worksheet

My measurable, time-bound performance goal (i.e., my desired outcome)

Things I want to start doing

Things I want to stop doing

Things I want to continue doing

Observable outcomes and indicators that I've made positive change happen

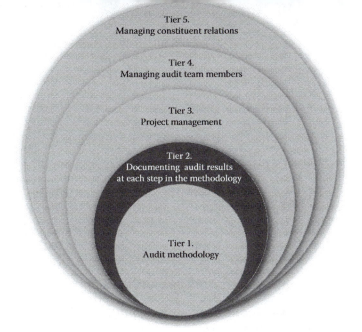

5 Documentation and Issue Development
The Building Blocks for Effective Audit Reports

CHAPTER CONTENTS AT A GLANCE

In this chapter, we will cover

- A general, four-step writing process
- Characteristics of effective audit documentation
- How to overcome writer's block
- Ways to organize your message
- How to develop audit issues based on test results
- Five elements of an effectively written audit issue
- Factors to consider when documenting audit concerns
- Ways to expedite report issuance

All writing is a process of elimination.

Martha Albrand

To save time and create efficiencies, many audit departments have created templates for recurring, routine correspondence such as documentation requests and engagement memos. These templates make it easier to communicate your thoughts because you only have to fill in a couple of blanks. However, we need to communicate other messages—ones that may be unflattering—in a manner that sparks agreement and action. I refer to this type of writing as summary (conclusionary or comprehensive) writing.

Summary writing is the product of critical thinking and is more challenging than completing a template. It is the expression of opinions and ideas based on review and analysis. During an audit, you will need to document

- Your ideas, suggestions, and recommendations concerning the audit's scope and objectives in planning memos
- The project's progress and problems in status reports
- The attributes of effective controls in narrative workpapers
- The meaning of the test results in test workpapers
- The overall effectiveness of the area or process's system of internal controls in reports

Each of these written messages is intended to prompt the reader's agreement and action.

Since report writing has been the sole topic of many books, we will examine audit writing from a different perspective. This approach will provide a framework for expediting report issuance and communicating your ideas in writing at any point in the audit, for example, audit planning results, detailed assessment outcomes, interim status reports, audit issues or concerns, and audit opinions. We will also focus on the elements of an effective audit issue, because well-written audit concerns or issues are the building blocks that make audit report writing easier.

We will not focus on audit report formats, because the layout and content of an audit report is the tangible manifestation of the chief audit executive's relationship with the board and C-Suite. Since each organization—including those in the same industry—has different organizational cultures, different risk management appetites, and different degrees of complexity, it would be presumptuous to recommend a one-size-fits-all report format. That said, there are specific behaviors that accelerate report issuance, and we'll review them in this chapter along with a four-step approach to drafting effective audit reports.

The documents we prepare during the audit are not haphazard collections of information. Our writing should convey the results of our careful thinking concerning the Critical Linkage™. Insofar as the Critical Linkage™ is a linear thought process, with one step building on the results and conclusions of the prior step, our writing should be logical and easy to follow. Readers should not require a specialization in internal auditing to understand the points we make—even when our points deal with the technical aspects of an operation, revenue recognition, or financial reporting.

Your workpapers document the audit work performed. They are the record of information obtained and analyses conducted during the course of an audit. Testing workpapers provide the support for the audit report, the basis for supervisory review of the work performed, and the means for the external parties to evaluate internal audit work. Your work at each audit phase must meet a reviewability standard. Specifically, your documentation must be able to stand on its own and allow another person to reach the same conclusion you did without asking you any questions or discussing the matter with you.

GENERAL, FOUR-STEP WRITING PROCESS

The Four-Step Writing Process provides a high-level view of the mental flow and activity that you should follow whenever you need to prepare written messages. It is helpful whenever you need to organize the messages you need to convey (Figure 5.1).

Determine the issue based on the analysis of test results

Organize your thoughts to express the message on the Issues List*

Write the report based on the Issues List content

Edit the report

Important: To save time, write all your audit documentation in a report-worthy manner.
* Issues List is a generic name to describe a repository for all concerns that emerge during an audit.

FIGURE 5.1   The four-step writing process.

Step 1: Determine the issue based on the analysis of test results. This is the result of data collection. Use the test workpapers to show concrete evidence of your findings. Make sure that you have responded to and resolved all of your project manager's review notes, that is, questions and comments. Figure 5.2, the Simplified Writing Process, will help you determine how to pace yourself.

Step 2: Organize your thoughts to express the message on the Issues List (a generic name to describe a repository for all concerns that emerge during an audit). List issues chronologically, if a time line is relevant. Group similar concerns to convey significance and pervasiveness.

Step 3: Write the report based on the Issues List content. Make sure your cross-references are accurate and complete.

Step 4: Edit the report. If you wrote in a report-worthy manner throughout the audit, this step shouldn't take long. You shouldn't be rewriting or inserting new information. You should be able to leverage existing documentation.

GENERAL CRITERIA FOR WRITTEN COMMUNICATION

The following are four ways to assess your writing's effectiveness:

1. It should prevent surprises. Consequently, it needs to be clear and timely.
2. It should be relevant to the business objectives' achievement and the reader's informational needs.
3. It should convey perspective and present a balanced view. Although our work and reports focus on identifying, analyzing, and concluding on exceptions, we need to acknowledge the positives as well.
4. It should be clear and easily understood by your reader. Consequently, jargon and technical terminology should be replaced with language that's relevant to and understood by the reader. Remember that your audience needs to
   a. Understand the situation
   b. Receive accurate and objective data in a document that is organized, clear, factually supported, objective, grammatically correct, balanced in perspective, timely, and relevant to their needs
   c. Make appropriate, informed, and justified decisions after reading the facts

DOCUMENTING THE AUDIT: GETTING STARTED

Admittedly, it is sometimes hard to get started. I have found it impossible to write until I know what I want to achieve and can envision the reader and understand his or her needs. Am I seeking approval, help, insight, or additional information? How experienced or knowledgeable is the reader concerning my topic? The answers to these fundamental questions enable me to determine the amount and type of detail to include.

At a minimum, I find it helpful to answer the following questions to determine how to organize my message:

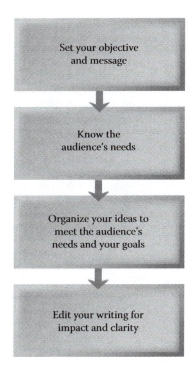

FIGURE 5.2   Simplified Writing Process.

- What is my message or point?
- Who are the readers?
- Why are they reading this document?
- What information do they need to understand the issues and the point I'm trying to make?
- Are my most important points expressed in the first sentences of the opening paragraph?
- Is my message expressed concisely, that is, in sentences of 25 words or less?

Now, you may be tempted to take the 25-word limit literally, and I would suggest that you not do that. The point of having a word limit is to sensitize you to a sentence's length. While shorter sentences are easier to read, long ones can be grammatically correct. If you write a long sentence, make sure that the subsequent one is short.

MIND-MAPPING APPROACH

Every so often I experience writer's block, and merely answering these questions is not enough. At other times, I am overwhelmed by the amount of information I need to cover. Sometimes I am just not sure where to begin. This is when I use mind mapping.

I am not sure who invented this approach, but it is a written form of individual brainstorming and requires a blank sheet of paper, a pencil, and a set of colored markers. It is useful when you are trying to organize your thoughts or determine themes in audit test results. Here's what you need to do to create a mind map:

1. View the blank sheet of unlined paper as a canvas on which you will record in pencil any thoughts, ideas, or issues you have regarding the process. Begin by writing the project's objective in the center of the sheet of paper.
2. Use the entire sheet of paper to record your thoughts. Do not try to record your thoughts in a linear or step-by-step manner at this point. Write your ideas on any available space on the page. Concentrate on using the entire sheet of paper to record all the ideas that come to mind. Do not evaluate your ideas at this time.
3. Once you have finished recording your thoughts, use the colored markers to group related thoughts together by color code; for example, red may be assigned to all thoughts that pertain to budget, and purple might be assigned to all thoughts that pertain to data-collection activities.
4. Once all your thoughts have been color coded, you are now ready to refine this information and convert it to a traditional outline format.

Although Microsoft Visio has mind-mapping symbols, it is easier and faster to create a mind map manually. Mind mapping requires the use of creative thinking, whereas using software is an analytical activity. When one tries to think creatively and analytically at the same time, writing becomes almost impossible.

"Stream of Consciousness" or "Don't Edit while Writing" Approach

Sometimes, mind mapping may not be useful because you are still formulating ideas. If you find yourself in this situation, you may want to begin writing the first thoughts that occur to you. Do not strive to write complete sentences at this point. You can develop paragraph and sentence structure later. The objective is to generate as many phrases as possible. When you use this approach, do not edit as you write.

Once you have generated as many written ideas as possible, review your list to convert the phrases into sentences. Next, determine whether any of the sentences relate to a common topic. If so, rewrite these sentences to form paragraphs. To complete your message, organize the paragraphs using a chronological, attention grabbing, rational, or devil's advocate approach.

Chronological Approach

- Cite the historic or past issues first.
- Cite the current situation.
- Cite the next steps or future situation.

Attention Grabbing Approach

- Get the audience's attention.

- Describe the audience's need.
- Satisfy this need, that is, describe the solution or recommendation.
- Ask the audience to take action.

RATIONAL APPROACH

- Describe a situation or example.
- Make your point.
- Explain the reason or rationale for your point.

DEVIL'S ADVOCATE APPROACH

- Make your point.
- Deliver a counterpoint or devil's advocacy position.
- Describe the argument.
- Draw conclusions.

Aspects of these strategies can be combined to form new ones. Your choice should satisfy your objective and the reader's needs. The effectiveness of a specific approach depends on your objective, the audience's needs and expectations, and the situation.

When communicating information concerning the audit, you will find using the chronological or rational approach to be the most useful.

Regardless of whether you use mind mapping, stream of consciousness, or something else to get started, remember that editing should be performed last.

TIPS FOR EFFECTIVE SUMMARY WRITING

Summary writing is the act of presenting data comprehensively. It enables you to communicate and present a wrap-up of facts and observations and of conclusions concerning this information.

It will be easier to determine your conclusions if your documentation is well organized. The characteristics of effective audit documentation are as follows:

1. Facts and findings are presented objectively and are quantified or supported by evidence.
2. Information is well organized to meet the reader's needs.
3. Abbreviations and acronyms are spelled out before they are abbreviated, so the reader understands the meaning.
4. The point of view, for example, the third person (i.e., "the Audit Department") or first-person plural (i.e., "we"), is consistent throughout the document. Don't mix the two perspectives.
5. The tense is consistent throughout the document.
6. Bulleted lists follow a consistent, parallel structure.
7. The sentence structure is grammatically correct and simple.
8. Dates are spelled out, for example, May 30, 2015, not 5/30/15.
9. Numbers are rounded to the nearest integer instead of using actual amounts when citing them in documents other than the test workpapers.

10. Et cetera (etc.) is used sparingly and only at the end of a list of more than three things. When your list has three or fewer things, use the conjunction "and".

11. Don't use the same word as a noun and a verb in the same or proximate sentences, for example, "The Operations Department's operational controls operate effectively." While it is grammatically correct to do so, it gives the reader the impression that you have a limited vocabulary.

12. Don't just write a conclusion, for example, that the area is being audited because it is high risk; provide some color to support it, for example, describe the nature and type of risk affecting the area.

To help you remember these points, keep the acronym CRAFTS in mind:

1. Clear: The documentation should be clear and concise. When writing, use simple, direct language and proper sentence structure. Do not use complex words or technical jargon as they may confuse or distract the reader and dilute your message. If technical words, phrases, or jargon must be used, be sure to define these terms early in the documentation. Strive to make your documentation easy to read. People are busy and want to obtain information quickly, concisely, and easily.

2. Relevant: Documentation should include only pertinent data; irrelevant information will only serve to frustrate and confuse the reader. Strive to present thorough, balanced, pertinent, and complete information. All sides of the issue should be reported fairly.

3. Accurate: All your documentation should properly reflect the facts. When writing, make sure you present the information in an unbiased and well-organized manner. Consider the reader's subject-matter expertise; this will help you decide whether to include background information.

   In addition to determining that your information is reported accurately, make sure your writing is grammatically correct. Documentation that contains spelling mistakes and grammatical errors will affect your credibility and undermine the audit. Since most computers have a spell check function, be sure to use it. Remember that the spell check will not verify contextual meaning. For example, spell check will not highlight "form" when you meant to write "from" unless the usage violated a grammar rule. Similarly, spell check will not highlight "manger" when you meant "manager." Before submitting anything, read it out loud for consistent use of tenses, proper sentence structure, and objectivity. Once you are satisfied with your message, make sure your format (use of margins, underlining, etc.) meets established standards.

4. Fair: When writing, remain objective. Avoid using subjective terminology and do not voice your personal opinion in audit documentation. Only facts and substantiated conclusions should appear in your documentation. Consider your tone, spin, and vocabulary—some words such as "fail" and "poor" have a negative connotation and trigger reader defensiveness.

5. Timely: Correspondence, documents, and reports that are issued long after the audit is completed have diminished value. If you remember to clearly document and write the test result conclusions and audit issues as you finish them, the report issuance process will proceed smoothly and rapidly. Well-documented workpapers make it easy to develop compelling, well-written audit concerns. Be sure to refer to your department's methodology and comply with its standards for report issuance, for example, the report should be issued X weeks after the closing conference occurs; and with the department's style guide.

6. Significant: Express facts, impact, and likelihood in a way that captures the impact and importance. The degree of importance and impact may change based on the target audience (e.g., managers of the area being audited versus executive level). For example, if you audited mailroom operations, you may identify concerns that represent high residual risk to the head of this area, that is, control gaps that threaten the achievement of the mailroom's business objectives. These same issues may have little consequence to the rest of the organization and would not be communicated to executive management in your organization.

ISSUE DEVELOPMENT

An issue is a description of the control weakness. Depending on your department's culture, an issue may also be called an observation, finding, or concern. Regardless of its name, this description includes the condition, that is, what you found as a result of the testing compared with what should be happening (the criteria or standard). An issue is typically a missing or nonfunctioning control. The issue's description needs to identify and explain why the deviations occurred; that is, it describes the root cause of the problem. The issue is generally expressed in the first one or two sentences of the audit observation (Figure 5.3).

Five Elements Every Audit Observation Must Contain: Element 1: The *criteria* or standard. This is a description of what the situation should be. It is the standard used to test the control, for example, the policy, performance norm, or guideline. The criteria may be implied if it were something a "prudent person" would do.

Element 2: The *condition* or situation. This is a description of what you observed or the results of your test, that is, *the facts in evidence*, quantified where possible. *Evidence* is the key term—it should be a description of what you found, that is, the test results supporting the issue. Evidence describes the exception or deficiency identified in the process, function, transaction, and so on. It can be any information needed to help support/justify the control weakness. Consider how the condition compares with the criteria or what was expected.

Element 3: The situation's *cause*. This is a description of the reason(s) why the exception occurred, preferably citing the root cause. The root cause is the *reason* for the difference between the expected and actual *conditions* found within an area, for example, lack of appropriate management control. The cause is cited if it is not self-evident.

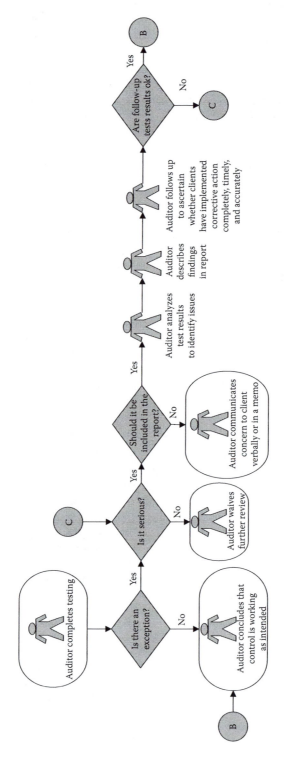

FIGURE 5.3　Life cycle of an audit issue.

Element 4: The consequence or *impact* that the exposure or risk poses concerning the achievement of the business objectives if the situation is not corrected. This provides details on the impact associated with the *condition* outlined in the issue. This element needs to answer the question, "So what? Why should a business manager address the cause of the situation?"

Now, if you are wondering how you will come up with the issue's impact, I suggest that you revisit the Risk Mental Model on page 46. When you used that model, you identified, analyzed, and assessed the inherent risks. You thought about what could go wrong to threaten the area or entity's business objective. Now that you have completed the testing, you know what went wrong; it is the condition compared with the criteria (what should have been happening). When you evaluated the inherent risk using the Risk Mental Model, you considered each risk's impact on the organization as a whole, that is, how much damage it could cause and whether this damage would have a financial, operational, reputational, or regulatory impact.

Element 5: The *corrective actions* that would either prevent the situation from occurring or mitigate the negative impact should the situation occur. The corrective action should specify who will do what by when. When writing this part, make sure you have verbally discussed the exception with line management. During this discussion, obtain line management's input concerning the corrective actions they plan to take, set target dates, and determine the responsible parties. Management's corrective action plans should be feasible and timebound, produce a measureable outcome, and address the root cause of the observation. Your role is to determine whether the plan meets these attributes.

If you followed the guidance for evaluating test results, you know that you executed the test correctly. You also considered the exception's possible causes. As part of this mental exercise, you might want to reflect on the section of the Risk Mental Model that relates to how or why the risk could occur to get some ideas before you speak with the area's management concerning the defect's root cause. Based on your understanding of the cause, you are able to help management come up with a useful corrective action plan and evaluate management's proposed corrective action to determine whether it will address the root cause of the control gap or breakdown and prevent it from recurring (or at least minimize its likelihood).

When Are Issues Found? Issues or potential issues may be discovered at any point in the audit process. For example, when you start the audit and are interviewing managers, they might bring issues to your attention. While documenting flowcharts or process flows during the control evaluation, you might determine that there are no controls or monitoring mechanisms over a significant risk. As another example, you might identify exceptions during testing. As issues arise, be sure to discuss them with the clients. Chapter 8 describes techniques for doing this.

Creating Common Terminology for Issue Development: During the testing phase, one or more things can happen. You can have no test exceptions or you may have a few or many. Some of these exceptions will be isolated incidents, recorded and explained in the test workpaper and communicated verbally to the process owner but excluded from the audit report. Other exceptions that initially seemed inconsequential when viewed as individual test results may be more serious when viewed comparatively or in the aggregate.

I think it's important to define and use consistent and specific terms to differentiate these exceptions. If your department has not created a lexicon, consider adopting the following one:

Audit Issue: One or more systemic, widespread and/or material exceptions that share a commonality, for example, cause, location, performer, activity. Issues may be communicated in the report (finding) or verbally or in a memo (observation).

Exception: An aberration or deviation from the anticipated results. Exceptions can be isolated, random, or systematic.

Finding: One or more related exceptions that are important and audit report worthy.

Isolated Exceptions: An aberration or deviation that is infrequent, a one-off, or insignificant.

Observation: One or more related exceptions that are important but not audit report worthy and are communicated to management by memo or verbally.

Test Results: The outcome of a test, which may consist of one, some, none, or all exceptions.

When trying to differentiate isolated exceptions from issues, a common reaction is to increase the sample size. This is not the most efficient practice. Instead, carefully review the test methodology to determine whether it is constructed in a useful, meaningful way. Then, consider the results of other audit tests to detect trends among results. Also, consider the control environment over the process under review. To determine importance, ask yourself, what is the risk to the business if this control isn't functioning and what is the impact or effect of the test results on the process? You answered this question in a slightly different way when you assessed the risks using the Risk Mental Model.

In order to analyze and distinguish observations from findings, it is important to have a repository for all exceptions identified during an audit. Depending on whether your department uses electronic workpapers and the software's feature functionality, you may already have this repository. If your department is using Word and Excel to document your results, consider creating a template to house emerging or existing issues.

An Efficient Way to Document Potential Issues: The following is an example of a format that can make it easy for you to document in one place all potential issues or concerns as they arise during the course of the audit. For our purposes, we'll refer to it as the Emerging Issues Matrix.

Assuming this document will be part of the audit workpapers, you should cross-reference all issues on the Emerging Issues Matrix to the source workpapers, for example, the test workpapers, so that anyone who reviews the work can locate the supporting documentation.

Creating a single document that contains a description of each exception makes it easier to remain organized when discussing these items with an constituent or audit reviewer. It also makes it easier to identify trends in the audit results that may exist across the areas or functions audited. Additionally, this document makes it easy to record the disposition of each item based on the results of discussions with constituents.

When using this document as a part of the workpapers, each potential exception must have a disposition, for example, is it an audit issue or not; and a reason for this disposition. Some examples of dispositions are

"Included in the report."
"Verbal comment only/minor concern."
"Resolved/waive further review."

All of the exceptions described on this document may or may not be reflected in the audit report, but they all will be communicated in some way with the area's management. If an issue is grouped with other similar exceptions, the auditor should note that on the document as well.

Table 5.1 and Figure 5.4 show two examples of the various layouts for this document.

GUIDELINES FOR IDENTIFYING ISSUES

As you evaluate your test results, the following actions will help you identify audit concerns (Figure 5.5):

- Make sure you executed the audit test correctly.
- Evaluate test results as you find them to confirm the accuracy of the tests and their results as well as to determine if additional testing is necessary.
- Be careful when evaluating test results. Do not apply the results of skewed or stratified samples to an entire population.
- Discuss the audit test results with the area or process's management to determine the cause of exceptions and audit issues.
- Keep an open mind. Do not jump to conclusions concerning causality.
- Vet assumptions concerning cause with the area's management to make sure your ideas are valid.
- Differentiate symptoms from causes. Symptoms are the test results that exist. Causes are the reasons why the test results occurred. You need to understand why the exceptions occurred so that management's corrective actions will be effective.
- Look for patterns or trends among the test exceptions. This will help you distinguish exceptions from audit issues.
- Look for trends among the results of several audit tests to obtain perspective concerning the area under review. Group similar concerns together.

TABLE 5.1

Documenting Emerging Issues: Option 1

Audit name: Indicates the name of the audit

Point Number	Issue Short Name	Issue	Issue Rating (H, M, L)	Resolution	Status
Numerical reference for each potential point or concern.	Brief title of the issue or point being documented. Examples include systems security–access, sales practices, inconsistent monitoring, and so on.	Provide specific information about the observation in this column. This description should include the condition, cause, criteria, and consequence. Describe potential issues found either through testing, risk analysis, or interviews. If the concern resulted from testing, include the following information related to the testing conducted: Population size from which samples were selected, sample size reviewed, number of exceptions noted, a description of the exceptions noted, and the impact or magnitude of the issue.	Indicates the auditor's evaluation of the rating of the issue identified. The logic used to determine this rating should correspond to the ranking associated with the risk the control is intended to mitigate.	Brief description of what the business group management team will do to address the issue. Information in this column should be the starting point for the action planning process. The action plan should address the root cause of the issue.	Identifies whether the issue is open or closed, reportable or nonreportable in the final audit report. If the item is to be combined with other points into an aggregated issue, this should be noted here and cross-referenced to all other points.

Documenting emerging issues: Option 2

Prepared By:	Date:	Index:
Reviewed:	Date:	Page:

Confidential: For discussion only

Audit issues and action plans matrix

Audit:
Report #:

ISSUE #	Condition and criteria	Root cause	Impact/ risk	Management response/ action plan/ responsible area/ target date	Status	WP ref
					__ Resolved – not an issue __ Verbal discussion point __ Written memo __ Audit report	
					__ Resolved – not an issue __ Verbal discussion point __ Written memo __ Audit report	
					__ Resolved – not an issue __ Verbal discussion point __ Written memo __ Audit report	

Guidelines for completing this form:

CONDITION AND CRITERIA – State the factual evidence that was found in the course of the audit. How does this compare to the criteria/what was expected?

IMPACT/EFFECT - Provide details on the impact associated with the *condition* outlined in the issue. What is the risk or exposure from the condition being different from the criteria?

ROOT CAUSE - Root cause of the issue. The root cause is the reason for the difference between the expected and actual *conditions*.

RECOMMENDATION - This is what Audit recommends to improve control in the environment. Recommendations are specific and address both immediate and future concerns related to the issue.

MANGEMENT RESPONSE/ACTION PLAN – This is management's proposed solution to the issue. It should include area that is responsible for ensuring it is resolved as well as a timeframe.

MITIGATING CONTROLS - This section is used to document any alternative controls within the unit or other units that might diminish the *effect* (exposure/risk).

STATUS – Document how the issue will be reported and disposed in the work papers. A standard selection will be available on the matrix as follows:

*LEGEND

Resolved – Not an Issue
Verbal Discussion Point
Written Memo
Audit Report

WP REF – Work papers that support the issue.

FIGURE 5.4 Audit Issues and Action Plans Matrix.

FIGURE 5.5   Audit and issue flow.

When looking for patterns and trends

- Look for commonalities:
  - Similar problems?
  - Similar flaws in control design or operation?
  - Similar causes?
  - Similar timing or time period?
  - Similar people involved?
  - Similar fix needed?
- Consider where the breakdowns occur:
  - Is it input processing?
  - Is it output processing?
  - Is it decision points?
  - Is it exception processing?
  - Is it monitoring?
  - Is it staffing, that is, incompatible functions?
  - Is it a deliberate act or oversight?
  - Is it an ignorant act or oversight?
  - Is it a cradle-to-grave issue?
- Focus on the risk issues:
  - Is it a completeness issue?
  - Is it an accuracy issue? (This could be recording, reporting, valuation, or measurement.)
  - Is it a legitimacy or authorization issue?
  - Is it a timeliness issue?
  - Is it an asset preservation issue?
  - Is it a people management or deployment issue?

Other Things to Consider When Evaluating Results: An audit issue, identified through audit testing, discussions, or analysis, is a deviation from accepted standards that impacts the business. Audit issues are identified by comparing what "should be" (i.e., expectations or standards) with "what is" (results of actual testing).

The following are some questions to ask yourself when considering the residual risk associated with the test results:

- Are the controls in place functioning as intended?
- Does the exception show a control breakdown or lack of control that may cause or has caused significant exposure to the business area?

- Could this control breakdown prevent the goals or objectives of the area from being met?
- Are there trends in the exceptions noted that point to a significant control weakness or pervasive problem (i.e., lack of monitoring, undefined control procedures, and inconsistent training)?
- Are there compensating controls in the area or another area that help mitigate the risk?
- What is the potential for negative synergy among missing or broken controls?
- Do the exceptions noted point to a control breakdown that is a repeat of a previous issue?

In addition, revisit the Risk Mental Model results and consider the type of risk that the control should mitigate, for example, financial, reputational, operational, technology (and its impact and likelihood).

You should also consider the number of transactions that will be effected and the dollar value of these transactions.

You should discuss all issues with the client as soon as possible and no later than the conclusion of the fieldwork phase of the audit. As soon as an issue is identified, discuss it with your audit team lead. Begin writing your issue immediately and share it with the client. This will help solidify the facts and allow the client to begin to develop corrective action plans.

EFFECTIVE AUDIT DOCUMENTATION TECHNIQUES

QUESTIONS TO ASK YOURSELF WHEN DOCUMENTING FINDINGS

- Did I clearly describe the exception or deficiency in simple terms?
- Did I explain the standard, that is, what should be happening?
- Have I clearly stated the impact (effect) to the enterprise?
- Have I used an appropriate tone?
- Have I quantified the issue or put it in perspective?
- Did I explain why the exception or deficiency occurred?
- Have I discussed my recommendations with the constituent?
- Has the action plan been worded so that it clearly indicates the constituent's agreement and ownership to implement the action step?

YOUR ROLE IN PREPARING AUDIT REPORTS

Let's face it—reading an audit report is not most people's idea of a relaxing read. But if the report is concise, well organized, and clearly addresses the issues, reading it won't be the focus anymore. The actions it inspires will be the focus of attention.

You can contribute to the timely preparation of audit reports by following these guidelines:

1. Prepare comprehensive, neat, and well-organized workpapers. This will make it easier for the audit team to support issues for inclusion in the audit

report. This means making sure that your cross-references are accurate and complete. It also means that you have responded to and resolved all of your project manager's review notes, that is, his or her questions and comments.

2. Make sure that you have documented each of the five elements of an issue.
3. Make sure that your writing is clear, concise, correct, complete, and appropriate in tone. The following questions will help you evaluate your own writing:
   a. Is it clear?
      i. Is the purpose stated clearly?
      ii. Will the readers know what response you expect from them?
      iii. Are the ideas organized in a way that would persuade the reader to accomplish your purpose?
      iv. Have you used language that the reader can readily understand?
      v. Does the language exclude technical terminology that would not be understood outside the area?
   b. Is it concise?
      i. Are unnecessary details excluded?
      ii. Have you eliminated unnecessary words?
   c. Is it correct?
      i. Is the information accurate?
      ii. Have you checked the spelling and grammar?
      iii. Is it written from a consistent perspective, for example, the third person (i.e., "the department") instead of the first-person plural ("we")?
   d. Is it complete?
      i. Are all the pertinent facts included? Make sure that your workpapers clearly and succinctly describe the audit issue and its cause and effect as well as the criteria used to evaluate the deviation.
      ii. Does it contain all the information the reader needs to make an informed, justifiable decision?
      iii. Are the exceptions' root causes identified?
      iv. Can the issue be quantified?
      v. Do management action plans include target dates?
   e. Is it appropriate in tone?
      i. Is the tone suitable for the reader's needs, the image you want to convey, and the context and form of your communication?
      ii. Is the action plan worded so that it clearly indicates management's agreement and ownership of implementation?
      iii. Is the content nonjudgmental, balanced, and objective?
4. Strive to write it right the first time. This is one of the most critical aspects of an efficient and effective writing process.
5. Expect revisions. Reports and other documentation are often reviewed and revised many times before they arrive at their final form.

Four Secrets to Accelerated Report Issuance

The following four actions will make it easier for you to draft audit reports quickly and competently.

1. Perform comprehensive and effective audit planning.
2. Understand and apply the critical linkage among the business objectives, risks, controls, and tests.
3. Write all audit documentation, beginning with the planning memo, in a report-worthy manner.
4. Sort the exceptions to identify the themes (e.g., by root cause, by process) you want to express in your report.

CHAPTER WRAP-UP

While it is imperative to document our conclusions, how you do this will depend on your organization's culture, complexity, sophistication, and risk appetite. These factors will affect

- Your audit report's layout and contents, that is, whether you report on all issues or just the repeat, high, and moderate ones
- Whether you provide an overarching rating to evaluate the entity or process, that is, satisfactory, needs improvement, or unsatisfactory
- Whether you provide a rating for each issue, for example, high residual risk, moderate residual risk, low residual risk
- Whether you make recommendations to close control gaps or only report management's corrective actions
- Whether you report on each finding as a stand-alone issue or analyze exceptions and amalgamate ones with similar causes or other commonalties to form issues
- Whether you document the condition, criteria, cause, consequence, and corrective action for each finding

In this chapter, we've reviewed general guidelines for documenting audit results and correctly identifying and documenting issues. Adopt the following four critical philosophies and let them drive your efforts:

1. The audit process is a logical thought process based on the Critical Linkage™ that functions as a building-block approach to report issuance. All effort should relate to and document the critical linkage among the business objectives, the inherent risks that threaten the goal achievement, and the controls that mitigate these risks.
2. All audit documentation should be written in a report-worthy style; that is, write it right the first time. Be sure to write conclusions on all the workpapers using a consistent writing style. This will make it easier to identify potential audit issues and complete the workpaper and file reviews.
3. Write for your reader's needs, not your own. Put yourself in the reader's position and anticipate the type of information and level of detail they will need to make a decision or take action. Convey this information as clearly, quickly, and concisely as possible.

4. Begin with the end in mind. Keep the audit report format, reader, and writing style in mind as you prepare your planning memo. Specifically, be sure to
   a. Write the audit objectives and scope in a way that you can easily cut and paste it into the audit report.
   b. Write the first paragraph or two of the area's background as though it were a news article. Eliminate extraneous information and concentrate on providing only those facts that a reader would need to size the area and appreciate the area's relevance vis-à-vis the rest of the organization.

If you keep these four points in mind when writing, you will find it easier to document your workpapers.

CHAPTER SUMMARY

- Use the four-step general writing process to organize your message.
- Use mind mapping or "stream of consciousness" writing to help you overcome writer's block.
- Use summary writing techniques to present your findings comprehensively.
- Make sure your audit observations contain five elements—the condition, criteria, cause, consequence, and corrective action.
- Group audit issues that have common themes or causes.
- Be sure management corrective actions target the root cause of the problem or control breakdown.
- All documentation should be written in a report-worthy style the first time.

QUIZ

1. Summary writing is
   a. The expression of opinions and ideas based on review and analysis.
   b. Documenting the information obtained.
   c. Filling out all fields in a template.
2. I should pay the most attention to writing
   a. When all issues have been identified and the testing documentation is compiled.
   b. Starting with the first workpaper I draft.
   c. When drafting the final audit report.
3. An issue is
   a. A description of what you found.
   b. A description of control weakness identifying the root cause of the problem.
   c. A description of the effect or consequence.
4. Well-written audit issues must contain
   a. Condition.
   b. Cause.
   c. Criteria.

    d. Consequence.
    e. Corrective actions.
    f. All of the above.
5. When identifying issues, avoid discussing them with the area or process' management.
    a. True—discussing may slant issue identification.
    b. False—this discussion will help determine the cause of exceptions and audit issues.
6. An effectively written audit report should
    a. Summarize the contents of control evaluations.
    b. Describe all test results.
    c. Trigger actions that resolve the issues raised.

The performance planning worksheet

My measurable, time-bound performance goal (i.e., my desired outcome)

Things I want to start doing

Things I want to stop doing

Things I want to continue doing

Observable outcomes and indicators that I've made positive change happen

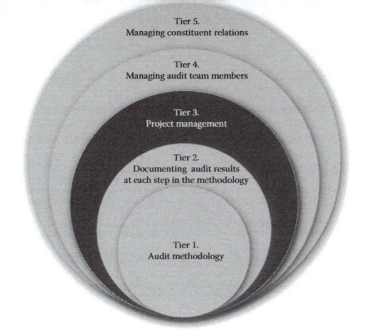

Tier 5.
Managing constituent relations

Tier 4.
Managing audit team members

Tier 3.
Project management

Tier 2.
Documenting audit results
at each step in the methodology

Tier 1.
Audit methodology

6 Core Competencies You Need as an Auditor

CHAPTER CONTENTS AT A GLANCE

This chapter will cover

- Why executive presence is a quality every auditor needs to demonstrate
- The importance of critical thinking and how to do it
- How to optimize your time
- How to set and manage priorities
- How to manage up
- How to prepare for and run effective meetings
- Ways to enhance your facilitation skill to encourage others to talk
- Tips for effective audit project reporting

We are what we repeatedly do; Excellence, therefore, is not an act, but a habit.

Aristotle

EXECUTIVE PRESENCE IN AUDIT

Executive presence is a popular term to describe a professional who exudes confidence, leadership, and trustworthiness. Emotional intelligence lies at the core of this quality. If you are aware of yourself and how you fit into the context of your job, you have laid the foundation for conveying executive presence. If you couple this self-awareness and self-management with an awareness of others and an ability to manage relationships with others, you have all the ingredients for an effective executive presence.

How do you know if you or someone else has executive presence? Outward signs of executive presence are expressed through

- Professional attire
- Posture
- Behavior
- Facial expressions
- Subject-matter expertise
- Fluency of speech (barring speech impediments)
- Personal values
- Initiative
- Demonstrated integrity

Why should you develop this quality? Your image telegraphs messages about you to others and contributes to their perception of you as a person and as a professional. While judging someone based on appearance seems unfair and superficial, it happens regularly. I remember preparing for my first job interview after graduating from college. I had absolutely no idea what I wanted to be. All I did know was that I didn't want to be unemployed. At that time, the job market was exceptionally tight. Jobs were incredibly hard to find. To increase my employment odds, I made an appointment with the career development office. Here's what I learned:

People form a judgment about you within the first 10 seconds of meeting you. Most people can't say their name to someone else and break the ice within 10 seconds! Everything you say or do after the first 10 seconds confirms or refutes the initial impression you made on the other person.

While you may quibble about the time frame for the formation of this first impression, it is hard to argue with typical human reactions. Once a person reaches a conclusion, this individual tends to focus on facts that reinforce the opinion that has been reached. In other words, people rationalize the decisions they make. If they decide that something is good, they tend to unconsciously focus on the data that supports this opinion. There's an anonymous story that supports this phenomenon. It's called "Good Luck Bad Luck!"

A farmer used an old horse to till his fields. One day, the horse escaped into the hills, and when the farmer's neighbors sympathized with the old man over his bad luck, the farmer replied, "Bad luck? Good luck? Who knows?" A week later, the horse returned with a herd of horses from the hills, and this time the neighbors congratulated the farmer on his good luck. His reply was, "Good luck? Bad luck? Who knows?"

Then, when the farmer's son was attempting to tame one of the wild horses, he fell off its back and broke his leg. Everyone thought this very bad luck. Not the farmer, whose only reaction was, "Bad luck? Good luck? Who knows?"

Some weeks later, the army marched into the village and conscripted every able-bodied youth they found there. When they saw the farmer's son with his broken leg, they let him off. Now was that good luck or bad luck? Who knows?

Everything that seems on the surface to be an evil may be a good in disguise. And everything that seems good on the surface may really be an evil. So we are wise when we leave it to God to decide what is good fortune and what misfortune, and thank him that all things turn out for good with those who love him.

Author Unknown

While the farmer in this story remained circumspect, those around him jumped to judge every event. This same behavior occurs frequently within our organizations. Impressions drive opinions, and soon the opinions become fact. Consequently, you need to consider and be responsible for the image that others have of you because it will drive their interactions with you.

So far, I've focused on selfish reasons for exuding executive presence, but there's another reason to develop executive presence. It is a desirable trait in future audit leaders. While it is convenient to think that leadership is a trait that occurs naturally—that is, that leaders are born, not made—the reality is that leadership needs cultivation. Part of leadership is executive presence. People are more inclined to follow people that project confidence and common values—two traits associated with executive presence.

While it is tempting to think of image as the sole attribute of executive presence, this would be a mistake as well as an oversimplification. Demonstrating executive presence requires more than simply looking good and projecting confidence. Knowledge and savvy are important components of executive presence. Those who display executive presence also demonstrate intelligence and knowledge. This knowledge may be technical or business related, but individuals with executive presence are able to contribute useful and relevant ideas.

Another important attribute of executive presence is the ability to mesh well with the organization's culture, that is, to be a good fit for the team. Part of this requires that the individual share and demonstrate the norms and values that the organization esteems. Many organizations, particularly the larger ones, communicate their core values on life-sized placards and signs posted prominently throughout the office. These values may be teamwork, collaboration, innovation, growth, service ... the possibilities are limitless.

When an individual's core values are aligned with the organization's core values, working relationships are inherently more harmonious, decision-making processes are smoother, and communication styles are similar. Conversely, when an individual's core values are out of sync with those of the organization, it can be painful experience.

When you are trying to mesh with your organization's culture, it is helpful to do a little thinking; specifically, critical thinking. You have to know what your organization values in order to leverage your skills and minimize weaknesses (or convert your weaknesses to assets by changing your behavior).

CRITICAL THINKING: WHAT IS IT?

Critical thinking is another important competency that effective auditors possess. Are you a critical thinker? To a degree, this is a trick question—it requires critical thinking to answer it! Even the most introspective of us is not 100% honest with him- or herself. What would your coworkers and boss say? The brutal truth is, if we're the only ones who think we're critical thinkers, chances are we're not. Since critical thinking is a skill, it can be cultivated and enhanced. If you love the attentiveness to detail that auditing requires, chances are you already possess the analytical traits that are fundamental to demonstrating critical thinking abilities.

What exactly is critical thinking? It means being able to develop strategic ideas, evaluate others' ideas, and maintain professional skepticism. It does not mean condemnatory fault-finding missions. Critical thinking is an essential skill and key success factor in the performance of audit work. Critical thinking is the art and science of absorbing, analyzing, synthesizing, and evaluating information and then determining what course of action, next question, or additional information is necessary to arrive at a clear understanding of the root risk, control (or lack of), or process. Critical thinking involves a healthy curiosity, mental discipline, and a professional skepticism to separate root causes from symptoms.

Critical thinking is a particularly important ability for all professionals because of the pressure to "do more with less" and the expanded use of technology and reengineered processes. Internal auditors require critical thinking in just about everything they do. It is used to identify and define the entities that will form the auditable universe. The auditor has to rely on critical thinking to determine the scope of each entity; simply creating a universe that mirrors the legal entity structure or organizational structure of the organization is not enough. Critical thinking is used to determine each risk-based audit's scope and objectives. During the planning effort, it is used to identify salient information that affects the nature of the audit effort. Critical thinking is required when assessing an entity's inherent risk as high, moderate, or low. It is used to evaluate control design and also to determine sample sizes and sampling methodologies. Critical thinking is also used to evaluate the significance of audit test results and to arrive at an opinion that reflects the control condition for the area under review. Simply put, critical thinking is a core competency for auditors at all levels.

Critical thinkers engage in empathetic, adaptive, and creative behaviors. They seek to understand the other person's point by keeping an open mind and demonstrating curiosity. When they state their understanding of the other person's point, their goal is to accurately translate the message along with the speaker's feelings and meaning. Their curiosity is real—it is ineffective (and perhaps impossible) to fake true curiosity. They demonstrate professional skepticism by verifying or corroborating the information they receive. They consider and identify the assumptions in the other person's message. They value nonverbal cues as much as verbal ones.

Critical thinkers are also adaptable. They are not hung up on organizational status, nor do they expect others to conform to rigid personal standards. If their client or colleague speaks imprecisely or unclearly, they attempt to extract the main ideas without demeaning them. They value innovation, carefully listen to and attribute ideas to the giver, and are reasonable. The Greek word for reasonableness, e·pi·ei·kes′, is often translated as "yieldingness." During an audit, yielding is important as long as it doesn't negatively affect the auditor's independence or distort facts. Conceding a point—especially a minor or inconsequential one—can go a long way in gaining the trust of the other party.

Critical thinking is linked to being conceptual and creative. Those who have mastered it convert negative, critical reactions into positive, productive, option-producing efforts. When confronted with a new situation, critical thinkers ask, "In what ways might I...." (IWWMI). IWWMI thinking stimulates internal brain-storming efforts. It means focusing on things they or others can do to address the situation.

WAYS TO INCREASE CRITICAL THINKING ABILITY

Critical thinkers aren't born—they are a result of consistently performing the following five activities:

1. Recognizing when you are making an assumption. For example, be able to support opinions and conclusions with demonstrable facts. You must also state your position by organizing and performing situation analysis to break down the problem into its components including root cause. When you perceive an error in logic, respond appropriately by being disciplined and tenacious in following a line of thought and questioning.
2. Demonstrating professional skepticism, especially when evaluating data and formulating conclusions. Question the source, completeness, accuracy and timing of available data. Use PNI (positive, negative, interesting) thinking to review and evaluate ideas before making decisions. Ask yourself
    a. What are the positives about the idea? Identify an aspect or perspective that is beneficial or good.
    b. What are the negatives? Identify an aspect or perspective that needs development or improvement.
    c. What is interesting about the idea? Identify an aspect or perspective that strikes you as appealing or different. Probe these areas and gather more information. Innovation stems from this exploratory approach.
3. Watching your language. Be alert to the use of words like "always," "never," "some," and "most." Psychologists tell us that these are no-no words because they overgeneralize the situation and create a defeatist setting for the conversation. A good rule of thumb is to watch out for over-simplification and reductionism; things are rarely that simple. Also, be wary of complexities because they can obfuscate; things are rarely that complicated.

4. Watching your language again: be alert to the use of the passive voice (which can camouflage or omit facts) versus the active voice. Passive sentences are more verbose and nondirective than active ones and do not convey any sense of urgency. There is no room for ambiguity in audits—your message needs to convey decisiveness, direction, and certainty. Do not use qualifying or conditional phrases like "seems to" or words like "minor" (unless there are major issues as well). State the facts and quantify whenever possible.

5. Differentiating fact from perception. Is it really or does it just seem to be? If multiple variables aren't measured independently, the conclusions are susceptible to the Halo Effect. First identified by U.S. psychologist Edward Thorndike in 1920, the Halo Effect is the tendency to make specific inferences on the basis of a general impression. If a company is doing well, with rising sales, high profits, and a sharply increasing stock price, the tendency is to infer that the company has a sound strategy, a visionary leader, and so on. Performance, good or bad, creates an overall impression—a halo—that shapes how we perceive its strategy, leaders, employees, culture, and other elements. If an area had clean audits in the past, there is a tendency to expect it to continue to have clean audits as well as a tendency to expect low or acceptable residual risk (a halo). If the data are full of halos, it doesn't matter how much we've gathered or how sophisticated our analysis appears to be. Performance is relative, not absolute. An organization can get better and fall further behind at the same time. The link between inputs and outcomes is tenuous. Bad outcomes don't always mean that managers made mistakes, and good outcomes don't always mean they acted brilliantly. Make sure you understand the critical linkage among the business objectives, business activities, risks, and controls. This will ground your approach in specifics.

The following are some additional ways to build your critical thinking capability:

- Understand your goal; that is, what do you want to achieve? What will be the takeaway or result? As you work, consider whether your actions and results are putting you on the path to achieve your desired outcomes. If not, why not? Address the reasons why your results are different than expected.
- Tap into your intellectual curiosity to motivate your learning. For example, when you think about the businesses or processes that you regularly audit, what do you want to know? Where is your understanding of what they do or why they do it a bit fuzzy? When collecting information, be sure to ask questions to satisfy the five Ws and one H, that is, who, what, when, where, why, and how.
- Be prepared to approach situations from multiple angles; for example, think about the situation as though you were the process owner, then consider it as if you were a customer, then consider it as though you were a competitor, then consider it as if you were the control performer, and

finally consider it as the auditor. Assuming these different perspectives will provide additional insight and understanding. This tactic is particularly useful when you really do not understand the rationale for some processes, functions or activities.

- Make sure that your mind is open to ideas but focused on your end goal. Frequently, there are several ways to achieve the same outcome.
- Think conceptually (big picture), not just tactically; that is, don't get so close to the trees that you can't see the forest. This is particularly difficult when you begin to evaluate the controls. Typically, the folks you interview are so involved in the control's execution that you succumb to the hallucination that the control activity that they perform is at the center of the universe and totally integral to the financial success of the entire organization!
- Look, pursue, drive for potential interrelationships. By itself, a condition may not appear relevant or significant; however, in conjunction with other factors it may take on a higher level of significance worth investigating.

Become your own devil's advocate. Make it a point to deliberately try to poke holes in the rationale for your own and others' decisions. Okay, so your family and friends will definitely find this behavior annoying, but it will help you anticipate objections and spot alternate solutions and additional ideas.

Now that you've developed and used critical thinking skills to align yourself to the organization, you should feel more comfortable approaching the audit.

GENERAL GUIDELINES FOR MANAGING YOUR TIME

Another critical skill is time management. Whether your organization has a predetermined time limit on audits or not, managing your time will reduce your stress and give you a sense of accomplishment. Thinking takes time. It's too bad that when we're thinking, it looks like we're doing nothing. As knowledge workers, we will have days when our thoughts are clear and productive. We will also have days when thinking will feel like we are running in mud. During these times, hours will pass and our computer screens will remain as blank as our minds, our fingers exhausted only by our unceasing write-delete-rewrite efforts. Unfortunately, audits are timebound projects that are insensitive to the timing and flow of our creative juices. While we are thinking critically, we need to make sure that we manage our time and priorities so that we're timely in producing results.

Following are some guidelines for you as a person, a professional, and a formal or informal team leader.

Guidelines for You, the Person: Demonstrate self-management to remain focused by

- Determining the three or four things that are most important to you
- Making time to reflect on your personal needs and goals
- Making time to reflect on your current progress and obstacles

Accept the need to adapt your schedule when necessary to accommodate changes in priorities that occur as you move through the different chapters of your life. There will be times when your career will be the most important thing in your life (typically when you are just out of school) and then there will be times when family is the most important thing (typically triggered by events like marriages, divorces, births, deaths, anniversaries, graduations, and other cycle-of-life actions). When these things happen, you may find that your first reaction is to try to broaden your shoulders and continue as though nothing has happened. If so, you will realize (at some point, hopefully sooner rather than later) that this is not a sustainable position.

Guidelines for You, the Professional:

- Clarify your understanding of the "big picture," including your company's mission and strategy. Understand your organization's competitive positioning, core values, and long-term objectives.
- Understand your organization's risk appetite and view of internal audit. Do they view internal audit as *the* organizational control? Do they view internal audit as a valued business partner? How aligned with the organization are your views as a professional?
- Understand your organization's priorities and where risk management fits within this landscape. Are risk management practices baked into daily operations or are they addressed in response to specific regulators or requirements?
- Understand the outlook for your industry. Is it growing, declining, or remaining steady? To what extent does or will the competitive landscape affect your organization?
- Focus on the economic trends that affect your industry. If you are in the transportation industry, the price of gasoline (in the short term) and the availability of alternative energy or new technology (in the long term) can affect your organization's ability to achieve its goals.

Guidelines for You and Your Team: Establish, communicate, and manage to clear, goal-focused performance expectations. Remember that your team members and others will be watching what you do much more than they will be listening to what you say, so make sure that your behavior mirrors your words. Communicate clearly and frequently to all constituents, so that all have the information needed to get results. Make sure that you take the time to communicate in person whenever possible. While e-mail is the prevalent and popular communication medium, capitalize on all opportunities to interact with others in person. This is the easiest way to establish and expand relationships with others in your organization.

The following actions will help you optimize your time and energy:

- Acknowledge when you are procrastinating and prepare for a generally negative reaction from others to this behavior. However, your procrastination may be a blessing in disguise. Sometimes procrastination is a positive.

To appreciate this difference, you need to understand when and why you procrastinate. Normally, I am an action- and results-oriented person. I typically start each day with at least one thing that I intend to accomplish. Usually, I have a list of "things to do." Consequently, when I have the same item on my "to do" list for several days or weeks, I now realize that this procrastination is a signal. This signal could mean any of the following:
- I don't know what I'm supposed to achieve; that is, the desired or end result is unclear or undefined. If this is the case, I contact the client to discuss performance expectations.
- I don't know how I'm supposed to achieve the result; that is, the steps or procedure are undefined or unclear. If this is the case, I begin by working backwards from the desired result to identify the things that need to occur.
- The job is distasteful, unpleasant, or boring (in which case I think about whether I can delegate or skip the activity). If I can't skip or transfer the activity, I promise myself a reward (or focus on the benefit I will experience) once the activity is completed.

- Know how you currently use your time and the time of day you are at your best, that is, morning or afternoons. This information will help you schedule your time to optimize your alertness.
- Articulate your priorities. This helps you to think through what needs to be done and in what order. Recognize that at various points in your life, your priorities will change. For example, when you graduate from college, your first priority is to get a job and an apartment. Once you get married, your priority shifts. And your priorities continue to shift as the chapters in your life are written. Each time your life enters a new chapter, you should articulate your priorities as a way of clarifying them and defining what the result should look like.
- Schedule no more than one or two key activities per day. Get comfortable with the reality that you will rarely complete anything in one sitting or without interruptions. Adopt short interval scheduling: plan your tasks in 20-minute blocks and don't plan to do more than two or three tasks per day. This will enable you to accommodate the unplanned interruptions.
- Consider the value of to-do lists and whether the list should be electronic or on paper. Either way, a list is useless if you do not look at it. I remember teaching a three-hour time management program for a client. Halfway through the program, a woman entered the room. Thinking that she was in the wrong place, I stopped the class to tell her that this was the time management class. She responded, "I know. I'm supposed to be here. Can you tell I need it?" Her honesty left me hysterical. Before I could say anything, as she took a seat around the table, she said, "I had this on my to-do list. I just never looked at the list." The takeaway: don't let this happen to you.

Segregate your list into major categories, for example, phone calls, writing, research (or short- and long-term projects), and analytical work. Each of these categories requires a particular and different skill set. When you jump between activities that require different skill sets—for example, trying to write a report while responding to an e-mail or while returning phone calls—you tire yourself out more quickly, often without fully completing any of the activities.

While it's important to manage yourself and your time, it is equally important to manage your boss. In fact, some might say that managing your boss and his or her expectations should be your first priority insofar as this person controls your professional mortality.

TIPS FOR MANAGING UP THE LADDER

As a boss, I know firsthand that bosses rarely like surprises, including surprise birthday celebrations, so savvy professionals learn to manage their boss's expectations. You need to be sure to let your manager know if there are potential problems or delays. Here are a few tips for managing up within the audit department or within your internal client's organization without feeling like a sycophant:

1. Begin your message with the end or goal in mind. This will help you to be concise. It will also help you focus on the objective.
2. If this is a follow-up communication, summarize the results of each prior conversation or meeting before proceeding to cover "new" material. This helps reestablish rapport and reiterates the major points that have already been discussed or decided.
3. As preliminary results emerge and you begin to draw initial conclusions, begin to discuss them informally with decision-makers and decision-influencers—preferably in person. Begin these discussions within the audit department before convening them with your internal constituents. Watch the participants' reactions and make appropriate adjustments to your word choice.
4. Use plain language. Expressions like "random sampling," "properly accounted for," and "weakness" will trigger suspicion and defensiveness in listeners. Technical terminology will provoke confusion, interruptions, and derailment.
5. Cover each topic slowly, using win/win terminology to explain the business impact of each issue or recommendation.
6. Use e-mail to confirm the outcomes and decisions reached during each discussion or meeting.

GENERAL WAYS TO ORGANIZE THE MESSAGE YOU WANT TO COMMUNICATE

Sometimes misunderstandings result in wasted time. This is why preparing your message and being able to run effective meetings are so important.

Begin by determining your desired results, because these objectives allow you to measure and evaluate your outcomes against a specific standard. Typical communication objectives are to entertain, persuade, or motivate an audience or to obtain, clarify, or give information.

The following are some strategies (or formats) for organizing your message. The effectiveness of a specific strategy depends on

- Your objective
- The audience's needs and expectations
- The situation

There will be times when you will have to provide an overview of the audit methodology, communicate status reports, or outline the actions that will be taken during a specific audit phase. When you need to do any of these things, you may find it helpful to use

- *The chronological approach*: This approach requires you to organize your ideas and message based on a time line, beginning with the first action taken and proceeding to each subsequent action in the order in which the actions occurred. This is effective when the history of a situation is essential to dealing with the current condition or future outcome.
- *Tell'em, tell'em, and tell'em again approach*: This admittedly redundant approach is probably the easiest one for listeners to understand. Begin by stating the purpose or goal of the presentation's topic. Essentially, you are telling the audience what will be covered. Next, cover the topic in sufficient detail based on your audience's needs. Finally, summarize what you have told them. Use this approach whenever you have key messages that you want to reinforce.
- *The outlining approach*: Begin by amassing your ideas and notes. As you review them, determine (1) whether the information should be included in the presentation and (2) where it should appear. Once this step is completed, examine the intended contents of each section of your presentation individually. Organize the information in each section. Plan how you will make transitions from one topic to another. This approach is useful when there is no inherent connection between your subtopics; that is, there is no chronological connection.
- *The index card approach*: As you collect information or identify key points you want to communicate, write each idea on its own index card. Once you've completed your research or finished your analysis, sort the cards to identify themes. Devote at least one paragraph to each theme.

Regardless of the approach you use to organize your thoughts, you need to

- Plan the wording you will use to transition from one topic to another: plan the questions you will ask to elicit feedback and participation from the audience. For example, after you have provided an explanation of the audit methodology during an opening meeting, you might want to ask the head of the area under review, "How does this approach compare to what you were expecting?" This open-ended question is intended to engage the other person and gain insight into his or her prior audit experience.

Have copies of handouts available for attendees, even if you have already e-mailed this information.

Once your message is organized, move on to organizing the meeting agenda. According to researchers Alexandra Luong and Steven Rogelberg, the frequency of meetings is more exhausting than the length (Cohen et al, 2011). So, make sure to organize your meetings to optimize efficiency and productivity.

PLANNING EFFECTIVE MEETINGS

The following are six steps to plan effective meetings:

Step 1: Determine your key points and goal, including the meeting flow and who will speak first, second, and so on.

Step 2: Identify your audience and where they will be, for example, phone, site, and so on.

Step 3: Outline your message.

Step 4: Prepare your notes and supporting documentation.

Step 5: Rehearse your message.

Step 6: Relax before delivering your message.

To make sure that you will get the most from your meeting, try asking yourself the following questions as you make your arrangements:

1. Will a meeting actually be useful and helpful in this case?
2. Are the right people going to be attending?
3. Has a clear agenda been prepared and distributed ahead of time?
4. Do you have all the supporting materials and supplies (i.e., any facts and figures gathered in preparation as well as markers, flipcharts, tape, etc.)?
5. Has everyone been informed of the meeting's time and location?
6. Have participants carried out any assignments needed to prepare for the meeting?
7. Do all members of the audit team have speaking roles during the meeting?

The acronym POSTAD TV helps you to remember how to plan effective meetings, particularly how to construct the meeting agenda, and then notify the meeting attendees:

Priorities, Outcomes, Sequence, Timings, Agenda, Date, Time, Venue

For the sequence of agenda items, put the less important issues at the top of the agenda, not the bottom. If you put them on the bottom, you may never get to them because you'll tend to spend all the time on the big issues.

Place any urgent issues up the agenda. Nonurgent items should be placed down the agenda—if you are going to miss any, you can more easily afford to miss these.

If possible, avoid putting heavy controversial items together. Vary the agenda to create changes in pace and intensity. Try to achieve a varied mix through the running order.

Be aware of the tendency for people to be at their most sensitive at the beginning of meetings, especially if there are attendees who are eager to stamp their presence on proceedings. For this reason, it can be helpful to schedule a particularly controversial issue later in the sequence, which gives people a chance to settle down and relax first and maybe get some of the sparring out of their systems over less significant items.

Also be mindful of the lull that generally affects people after lunch, so try to avoid scheduling the most boring agenda item at this time; instead, get people participating and involved after lunch.

BUILDING YOUR FACILITATION SKILLS

Develop a reputation for running productive meetings and discussions. This means that you will need to hone your facilitation skills.

Facilitation is the process of encouraging others to communicate and exchange ideas. It is a skill that enables groups to work cooperatively and effectively. It enables a meeting or educational session to flow more smoothly; "to facilitate" means "to make easier." Facilitation is particularly important in circumstances where people of diverse backgrounds, interests, and capabilities work together.

A facilitator

- Encourages participation
- Maintains focus on the task
- Helps build small agreements
- Manages the group's process of decision-making

To be an effective facilitator,

- Create an open environment by breaking the ice and setting an upbeat tone for the meeting.
- Make individuals feel comfortable to express their ideas.
- Encourage all participants to listen to what others are saying. If a session is splintering into separate discussion groups, halt them politely and ask them to deal with one discussion at a time.
- Involve all participants. In any group, some individuals will be less inclined to speak up. Some people may be so quietly spoken that they are susceptible to interruptions by others.
- Watch out for signs that people are not involved. Be aware of participants keeping their heads down, doodling, or showing other symptoms of disengagement.
- To engage and encourage these people, ask them for their opinions and comments.
- Pay attention. As facilitator, you must be attentive to what is happening at all times. Do not get sidetracked into long discussions with individuals.
- Lead by example. You can encourage cooperative behavior by behaving in a way that is at all times honest, open, respectful, and nonpartisan. If

a disagreement arises, do not take sides. Instead, summarize the facts and the source of the disagreement. Then ask the group to resolve the issue.

There are different types of questions to facilitate discussions: If you want to acquire more data, ask *fact questions* like

- What happened?
- Who did/said...?
- What are the steps?

If you want to understand others' opinions and perspectives, ask *reaction questions* like

- What is your reaction?
- Do you agree/disagree? Why?
- What is challenging about the process?
- What questions does this stimulate in your mind?
- What needs more clarification?

If you want to understand causality or evaluate a situation, ask *analysis questions* like

- Why did they do...?
- What is the real source of conflict?
- Who seems to be most at fault?

If you want to spark thinking about the ways people can use the information under discussion, ask *application questions* like

- What are the implications for you in your job?
- How does it compare to your world?
- What can you use tomorrow at work?

REPORTING ON AN AUDIT'S STATUS

Status reports provide you with an efficient way to communicate the project's progress and any potential problems. Depending on the nature of the projects, it may be helpful to issue status reports weekly, biweekly, or monthly. The following is a suggested format for status reports:

Accomplishments described in bullet format. This section should summarize your results and conclusions.

Potential problems or issues described in bullet format. This section provides an early warning that a delay may occur. This section should also include a description of your contingency plans and/or options.

Next steps to be taken to complete the rest of the project plan, also described in bullet format. This section describes the immediate steps you will take during the next reporting period.

Do's and Don'ts for Project Status Reporting:

- Do put yourself in the reader's position and provide sufficient detail to answer the core questions: who, what, where, why, when, and how.
- Don't just list the tasks you've accomplished—this is evident by reviewing your updated project plan. Include a description of your results and conclusions.
- Do keep a written record of significant conversations and meetings you have had with others during your project's life cycle, especially when the meeting or conversation resulted in a decision. Having written records compensates for selective memories and makes it easier to recall who agreed to do what when. Don't rely on your memory.
- Do attach to your project plan copies of pertinent interim reports or findings—ones that support or expand on the accomplishments and potential problems you've cited.
- Do publish the results of data analysis, including your conclusions and recommendations. This will help you prepare your final presentations to management more rapidly.
- Do explain in writing why potential problems exist so that the best and long lasting solutions can be developed. If you don't know why the problem exists, continue exploring the issue until you do.

In addition to providing audit management with status reports, it is beneficial to have regular status meetings with line management during the course of the audit and especially during the testing phase. This is your opportunity to discuss findings as they occur.

Clearly communicate those test results that indicate an effective control design as well deficiencies identified through testing and those that may still require validation either by the audit team or verification/feedback responses from management.

Some talking points to cover during these meetings:

- Thank constituents and their staff for their cooperation.
- Review the nature of the work you performed because they may not be aware of its comprehensiveness.
- Present your findings in priority order and tailor the level of detail to the audience.
- Define the benefits of addressing the risk or control breakdown by thinking through the impact if the situation continues unaddressed.
- Get management's commitment as to when a corrective action plan will be developed, when appropriate.
- Explain the next steps in the audit process.

CHAPTER WRAP-UP

The way you, as an individual, approach an audit (whether you are part of a team or not) greatly influences its success. Enhancing your executive presence and critical thinking skills will help you anticipate and deflect personality clashes and misunderstandings if they arise. It will also help you during meetings with constituents and audit management. In my experience, internal auditing can be highly stressful. You will greatly reduce stress if you manage your time well and maintain open and effective communication.

CHAPTER SUMMARY

- Executive presence = self-awareness + self-management × awareness of others + ability to manage relationships
- Critical thinking = the ability to develop strategic ideas, evaluate others' ideas, and maintain professional skepticism
- Time management = Managing your time, your priorities, and your boss's expectations to produce timely results
- POSTAD TV, an acronym for planning effective meetings = priorities, outcomes, sequence, timings, agenda, date, time, venue

QUIZ

1. Executive presence is about the image we portray.
   a. True
   b. False—it also includes innate qualities like knowledge and intelligence.
2. Critical thinking is not crucial if you are not leading an audit.
   a. True
   b. False
3. Self-management involves
   a. Determining the three or four things that are most important to you.
   b. Making time to reflect on your personal needs and goals.
   c. Making time to reflect on your current progress and obstacles.
   d. All of the above.
4. When preparing a meeting agenda, less important issues should
   a. Be excluded because the risk is insignificant.
   b. Go at the top of the agenda so that they don't get eclipsed by the bigger items.
   c. Be sent in an e-mail after the meeting.

The performance planning worksheet

My measurable, time-bound performance goal (i.e., my desired outcome)

Things I want to start doing

Things I want to stop doing

Things I want to continue doing

Observable outcomes and indicators that I've made positive change happen

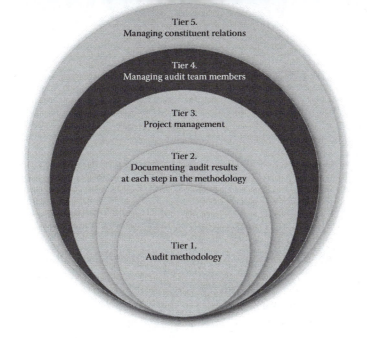

Tier 5.
Managing constituent relations

Tier 4.
Managing audit team members

Tier 3.
Project management

Tier 2.
Documenting audit results
at each step in the methodology

Tier 1.
Audit methodology

7 Techniques for Managing the Audit Team

CHAPTER CONTENTS AT A GLANCE

This chapter will cover

- The attributes of an effective team
- How to give tactful feedback
- The difference between leaders and managers
- The source of organizational vision
- The value of setting objectives and expectations
- Considerations when selecting the right leadership style for your team
- Practical ways to build a team
- How to delegate without worry
- How to motivate without money
- How to create organizational change

If you want to go fast, go alone. If you want to go far, go together.

African proverb

If you are the only auditor in your organization, you may rely on the work efforts of externals to help you complete the annual audit plan. To oversee this work, you will need managerial ability. The rest of the time, as you work on assignments by yourself, you will need leadership ability to influence and direct the audit project and to get others to agree and buy in to your findings.

If you are currently in an audit staff role or working in a one-person audit department, you might be inclined to skip this chapter because you don't think the topic is relevant to your situation. If so, you may miss an opportunity. Staff auditors who perform effectively get promoted and require leadership and managerial skill. Scanning the contents of this chapter will give you a preview concerning the issues you will have to handle at some point in your career.

Generally, the staff auditors selected to fill project leader roles or senior auditor positions have been high performers. They were able to accomplish prodigious amounts of quality work while eliciting few review notes. Their work papers were consistently well organized and written in a manner that enabled the work to "stand alone." Effectively, these auditors, who demonstrated ability as accomplished individual contributors, are selected from their peers to manage and lead the audit projects—often with little or no formal training. Up to this point, they have been apprentices, watching the behaviors of their project leads and emulating what they observe. If they were fortunate to have effective role models, they have an advantage as they lead their jobs. If not, well ... the experience will be bumpy for all involved.

Typically, new project leads focus on a singular goal of accomplishing the audit, as they should. However, they give little consideration to the interpersonal aspect of the project. They believe that the traits and behaviors that they displayed as staff auditors—the ones to which they attribute their promotion—are the ones that every auditor should display. Essentially, it's common for new project leads to expect others to behave as they do. When this does happen, the project lead misses an opportunity to hone his or her managerial and leadership styles. When this doesn't happen, disagreements can occur and, if unresolved, can pave the way for conflict.

Before going any further, I find it helpful to begin with the end in mind. In this case, let's begin by defining the attributes of an effective team. Before reading about these characteristics, think about effective teams that you either have been a part of or know about. If you were auditing a group of people, what attributes would you want to see that would indicate the members of this group were a team and not just a collection of individuals? Following are the attributes of an effective team. As you look over this list of characteristics, consider how your team measures up.

1. *Crystal-clear goals*: The nature of the work to be performed by the team is clearly communicated to and understood by the team members. However, before the nature of the work can be communicated, the leader needs to create the goal and make sure that it is well thought out and clear. The time-tested acronym SMART (Doran, 1981) is frequently associated with the characteristics of effective goals. The SMART acronym stands for specific, measurable, actionable, relevant, and timebound. To appreciate the value of incorporating each of these characteristics in a goal statement, compare the following two goals:

a. To audit effectively and complete the entire audit plan

b. To complete 100% of the annual audit plan before year end while addressing management's special concerns and without sacrificing quality

While the first goal appears clear, if we needed to achieve it, we would soon start to wonder just what it means to engage in "effective auditing." And, arguably, one could complete an annual audit plan by doing drive-by audits, that is, narrowing the scope or the depth of the testing in order to complete the audit within a relatively short amount of time.

While it's tempting to come up with goals like "perform effective audits," consider how these types of goals will be interpreted by the rest of the team. If the goal is not SMART, individuals could work on projects that are not relevant to the rest of the organization or meaningful in view of the organization's strategic plans. If your goal is not measurable, you will not be able to quantify your results and make year-over-year performance comparisons. You won't be able to answer the basic questions concerning your team's capacity to perform, that is, the estimated number of reviews you can complete in a year.

2. *Role clarity*: Each team member understands the part she/he has to play in order to achieve the goals. They understand their authority limits. Team members understand and respect the interdependency of their roles; that is, if one team member does not perform his or her role, the consequences for the rest of the team are known. Team members understand, appreciate, and capitalize on the strengths each member contributes to the team. This begins with an understanding of one's job description and transcends the traditional organizational chart. When team members understand their interdependency, each team member is able to anticipate and respond to the work demands created by unforeseen events, for example, illness or delays in receipt of constituent-supplied information.

3. *Effective leadership*: The team members receive adequate direction, oversight, and control. The team leader selects and uses a leadership style that is appropriate for the nature of the team goals and the background and experience of the individuals who comprise the team. Later in this chapter, we will explore your choices in selecting a leadership style as well as your choices in managerial style. In short, though, your style needs to adapt and change depending on the composition of your team, the types of audits your team needs to accomplish, the experience of the people, and the stability of the processes your team audits. Your challenge is being able to adjust your leadership style as needed to these different variables.

4. *Effective project management*: The team uses a consistent, logical methodology to plan, direct, manage, control, and report on accomplishments and potential concerns in order to achieve its objectives within predetermined time frames and constraints. The team uses a sound, consistent process or methodology to accomplish tasks or activities within time, cost,

and operational constraints. To the extent that your methodology is clear, cogent, and well documented, your team members should be able to follow it easily, and your internal quality-control reviews should be able to identify and correct behavioral deviations.

5. *Effective communication*: Team members exchange information openly and in a timely manner. This means that they leverage the right medium at the right time—for example, in-person discussions, text messaging, and e-mails—and are able to use the right combination of media to deliver the message, build the relationship, and maintain a record of key decisions and facts. Team members feel comfortable discussing real or perceived problems without fear of retribution or reprisals. They are effective at using consensus-building, group problem-solving, and influencing techniques to reach decisions. Being able to deliver feedback is a big part of effective communication. This feedback may be intended to reinforce or encourage the repetition of specific, desirable behaviors or it may be intended to extinguish undesirable behaviors. Typically, auditors use review comments as a means of delivering feedback. Specific tips for delivering effective feedback are included in this chapter.

6. *Managed conflict*: Team members appreciate and respect each other's personality and work-style differences. Differences are identified, addressed, and resolved in a manner that enables all team members to feel valued and to achieve their objectives. Effective methods of dealing with conflict enable both parties to exchange perspectives without judgment, name calling, and interruption. The root or basis for the disagreement is articulated and addressed with a view toward resolution. While it is tempting to ignore a disagreement, hoping it will go away, when it recurs (and it will recur if it is a fundamental or essential difference), a more assertive and proactive method is needed to resolve the brewing conflict. When this does not happen, that is, when managers ignore the situation or behave in passive-aggressive ways by overrelying on sarcasm or snarkiness, the effect on all team members is debilitating.

DELIVERING FEEDBACK

It's not enough to have a goal; you have to monitor performance to make sure that you and your team are heading in the right direction and spending time on activities that will result in goal achievement. Feedback is one of the ways to do this, and it is probably the most powerful method you can use to shape each team member's individual performance. Make it a point to provide feedback to individuals who are producing wonderful results; they need to know that their work is appreciated and they need to know exactly what you liked about their work so they can repeat this behavior on demand to give you top-notch results. It is equally important to provide feedback to those whose work is not satisfying your performance expectations. Your feedback is your primary way to shape your team members' behavior.

I used to dread having to give my team members feedback because I wasn't sure where to start or what to say. Then I realized that most people want to do a good job; the issue is defining what "good" looks like. Most team members just want a focused message that describes the specific things they either need to do, continue to do, or stop doing, supplemented with some examples.

Developmental feedback is given when behavioral change is needed. The key to delivering constructive developmental and reinforcement feedback is to use TACT:

- Tell: Talk about the other person's behavior.
- Affect: Describe how the behavior affects you or the organization.
- Change: Request a change in the behavior.
- Trade-off: State the positive consequences of a change in the individual's behavior.

Be prepared to deliver the "tell" and "affect" parts of this model together in one or two sentences.

Tell: When describing the "tell" part of the message, describe specific instances and occurrences. If possible, have copies of the work product so that the conversation stays focused on facts and not feelings.

Describe the undesirable behavior—the specific action and/or statement—to your team member. Don't tell her what she doesn't do; stick to what she does do that bothers you. Discuss the unwanted action without voicing your feelings about it. Behavior is objective; feelings are subjective. Also, mention the frequency of the behavior. Don't use words like "always" or "never." If you do, your team member is likely to become defensive and tell you about the one time she didn't behave in the way you have just described. Always be calm and nonjudgmental. Address only one unwanted behavior at a time.

Affect: Be sure to express how the troublesome behavior affects the organization, not what you think about your team member or her behavior. Be low key, not dramatic. You are negotiating a change in behavior, not trying to make her feel bad. When describing how the performer's behavior affects the rest of the department, do not exaggerate. Simply explain the impact. The point is to help the performer understand that his or her actions have ramifications—ones that this individual may not be aware of.

If the individual's behavior is exactly what you want, that is, this person is delivering terrific results, you want the individual to know this and to continue to do whatever it is he or she is doing. This is a time to deliver reinforcement feedback. To do this simply share the "tell" and "affect" segments of this model and then say "Thanks, keep it up!"

If you want the individual to change behavior, then you need to continue and deliver the "change" and "trade-off" parts of the model.

Change: Tell your team member what you want her to do. Limit your request to one or two specific actions. Don't be overbearing or dictatorial. Ask for feedback and agreement concerning the type of change needed. Perhaps you will need to change your behavior in this situation.

When describing the "change" part of the message, describe behavioral changes that the individual is capable of making. For instance, asking a short person to be taller or a tall person to be shorter won't work. Be prepared to deliver the "change" and "trade-off" parts of the model in one or two sentences. When delivering the "trade-off" part of the message, leverage your knowledge of the individual and the things that motivate him or her.

Trade-off: State what your team member is apt to gain by changing her behavior. Think carefully about what she wants and what will motivate her, keeping in mind that people are more apt to be motivated by reward than punishment. If you have spoken with this person about this problem before and you think it's necessary to mention negative consequences, then do so—but try a TACTful message first. Avoid threats; they only lead to counterthreats and arguments. Be careful not to use judgments and labels. For example, don't say "you're stupid or disorganized or wrong," and so on. Judgment words reflect opinions; stick to the facts.

The following chart compares two ways to give developmental feedback to a team member.

	Poor	Good
Tell about the behavior	You are constantly leaving priority work to help Muffy. That's ridiculous because you're falling behind in your own work.	I've seen you at Muffy's desk three times today.
Affecting the organization or team	You're pampering Muffy too much.	I'm concerned because your work is suffering. Your last report was late and incomplete.
Change in behavior needed	Stay out of Muffy's office. She'll figure out her assignments if you leave her alone.	I think you would do better to help Muffy once or twice a day at most. What do you think?
Trade-off in terms that benefit the performer	If you don't, your work will continue to suffer and so will your chances for a promotion.	Your work has been good up until now. If you concentrate on your own assignments, you'll be able to improve—and that means a better job review.

TIPS FOR PREPARING TO DELIVER TACTFUL FEEDBACK

Prior to delivering feedback, collect your thoughts and plan what you want to say during a coaching session. Give special attention to your word choice. At the end of the session, you know you have been successful if the other parties remembers what they said concerning their development plans more than they remember your words describing the need to change. If the other parties leave the coaching session with your words echoing in his or her ears, your wording may have been too harsh or direct.

	What you want to say:
Tell about the behavior Talk about the performer's behavior. What examples do you have to support your point of view?	Describe the team member's actual behaviors and cite specific instances. Be prepared to review examples of the individual's work. Focus on the actual things the performer is doing or saying. These behaviors are either positive (desired) or negative (undesired). Or, describe the behaviors needed from but not demonstrated by the team member, for example, using nontechnical language or remaining calm. Whenever possible, describe the reason or cause for this behavior, if you know it. Otherwise, the easiest way to determine the cause is to ask the performer to tell you why he or she is or isn't performing in a certain way. Sometimes the cause is just a simple misunderstanding concerning what the desired behaviors should be.
Affecting the organization or team Describe how the behavior affects you or the organization.	Describe how the performer's behavior affects customer relations, the rest of the team, and the organization.
Change in behavior needed Request a change in the behavior. Think about the steps the performer needs to take to make the change. Remember that the action plan should be realistic and address who will do what by when.	Describe the desired behaviors you want to see the team member demonstrate on a regular basis. This is a description of the actions he or she will take to either continue the desired behaviors or change the undesirable ones. *Be open to ideas the performer may have concerning how to make the change.*
Trade-off in terms that benefit the performer State how the change will have positive consequences for the performer.	Explain how the team member will benefit from making the change. This is one of the most important steps in delivering TACTful feedback. In order to come up with benefits, think about the things that are of value to the performer, for example, recognition or autonomy in scheduling or access to senior management for networking.

It's better to deliver feedback sooner rather than later, particularly when the feedback is developmental and you really want the individual to change his or her behavior. If you decide to give feedback sooner rather than later, you will find that the delivery is easier because you haven't allowed bad feelings to build over time.

The real question is, how soon is soon? Should you deliver feedback the very first time the individual does something that you don't like or doesn't meet performance expectations? Generally, I would advise you to speak out and deliver the TACTful feedback message immediately. However, in reality, I think your timing will depend on the situation and the individual. All gaps in technical performance should be recorded in review notes so that the individual may respond in writing to clear them and in so doing create stand-alone workpapers.

Other behaviors that benefit from feedback are not always so clear cut, for example, tips for running meetings effectively. For these situations, I think that you should set a predetermined limit of times you will allow a performer to do something you find objectionable before you deliver feedback. Insofar as you want to develop a reputation for being fair, you don't want to appear to jump on someone the first time they do something you don't like (assuming, of course, that the individual's transgression is not critical). You want to give the individual time to self-assess and self-correct before providing verbal feedback.

TIPS FOR MANAGING A MULTIGENERATIONAL WORKFORCE

As a result of healthier lifestyles, there are five generations in the workforce. Effective leaders will not only need to become familiar with each generation's core values and workplace habits, they will also need to learn to create symbiotic relationships, playing each generation's best qualities off those of the others to leverage and capitalize on workforce diversity. If you are in an audit leadership position, your leadership style is a pivotal factor because it affects the culture and tone within the department. It affects your ability to attract and retain talented auditors.

Your leadership and managerial style also affects your department's image within the organization, because the people who report to you are your emissaries within the organization. Essentially, they are your representatives. If they are listless or demotivated because you are not using an effective leadership and managerial style, this will affect their relationships with their constituents.

I don't know if you have ever considered the truth in the adage "People join organizations, but they leave bosses." Think about people you know who have accepted new jobs. Reflect on the conversations you had with them when they started the new job. Typically, the new hires brag about their new employer's organization as a whole, the scope and nature of the job they will perform, or the reduced commute they will have. Meet up with the new hire after he or she has been on the job for at least six months and, if things are not going well, the person's tune changes as does the focus of the conversation. This person's boss is now the conversational focal point.

LEADERS AND MANAGERS

The key to developing a high-performance multigenerational audit team is to be an effective leader as well as an effective manager. Not all leaders are managers, and not all managers are leaders. "Manager" is a title given to anyone who directs, coordinates, and controls a group's efforts to achieve an organization's goals and objectives. If you are at or above the project leader or senior auditor role, you are de facto in a managerial role. The critical issue is the effectiveness and appropriateness of your leadership and managerial styles.

Whenever I ask people in a classroom setting to describe the differences between managers and leaders, their responses indicate that leaders are more interesting, exciting, and compelling than managers, who are characterized as executors and maintainers.

When I bring to the participants' attention that their responses have tended to deify leaders and denigrate managers, they pause but are still not sure how to express the difference that they clearly have perceived and experienced.

Both leaders and managers demonstrate an ability to work with others, handle conflict, run meetings, and set priorities. However, leaders distinguish themselves from managers in the following ways:

- Leaders manifest calculated risk-taking and results-oriented behavior. They are comfortable taking risks because they either like certain types of stress or deal with stress effectively.
- Leaders are able to see problems where others don't and to arrive at solutions that others don't see.
- Leaders are persistent, self-confident, and have a distinct sense of personal identity.
- Leaders are able to influence other people's behaviors.
- Leaders identify with superiors and organizational goals.
- Leaders accept and deal effectively with a wide range of personalities. They are adaptive to myriad social and business situations, people, and circumstances.
- Leaders understand the necessity of keeping pace with new ideas and technological developments.
- Leaders have a well-developed network of contacts in and out of their organization.

The term "leader" implies setting direction. To do this, one needs an external orientation and focus to determine a course of action despite the future's inherent uncertainty. In contrast, the term "manager" implies handling or taking care of something. Effective leaders may or may not be in managerial roles. Effective leadership does not require position authority, that is, a title. This is known as informal leadership, and it only requires the ability to inspire others to follow. Depending on the effectiveness of the team's appointed leader, that is, the person with the title, the informal leader may be more influential and could potentially undermine the appointed leader's authority.

Both managers and leaders identify with superiors and organizational goals. Both managers and leaders accept and deal effectively with a wide range of personalities. They are adaptive to myriad social and business situations, people, and circumstances.

While leadership and managerial ability are both required, some people display an inherent preference of one over the other. Some individuals prefer situations that require them to take a situation, team, or department from a chaotic or problematic state and create solutions. Literally, these individuals thrive on instituting organizational change, for example, establishing an internal audit function for the first time. To tackle an endeavor like this requires what I call a "design and build" orientation, which is characterized by an external awareness and orientation. Once the new internal audit department is up and running, the typical "design and build" professional gets a little bored and is eager to move on to the next "hair on fire" situation that requires transformation and change. In contrast, other individuals are enthralled with the challenge of nurturing and developing the skill sets and abilities of their team members. These professionals see the potential in each of their team members and work tirelessly to tap into it and develop it. I call these folks "nurturing developers," and they naturally display an internal focus or orientation. Inherently, the design-and-build folks display more leadership ability while the nurturing developers display more managerial ability. However, both are needed. A design-and-builder without managerial ability will attract team members initially but will burn them out over time and starve them of recognition and a sense of completion and accomplishment. Similarly, a nurturing developer without an appetite for transformative change will create a wonderful esprit du corps that mushes along and tends to the damn dailies. Over time, those team members who have career aspirations will leave because the department isn't going anywhere ... It's a nice place but trapped in a performance rut.

Your challenge as a leader is to keep all of your team members—regardless of their generation—focused, energized, and highly productive, while defining the criteria that define the attributes of a quality audit work product. A crucial step is to empower your employees to appreciate and leverage other generations' workplace habits. This is even more critical in today's competitive environment, in which our organizations need marathoners, not sprinters, if we are going to win the endurance competition triggered by the global economy.

Although I have been fascinated by generational differences for decades, I wonder whether what comes across as generational difference is actually a cycle-of-life characteristic. For example, individuals starting out in their careers—especially those who have just graduated from college—tend to have more energy and optimism than those nearing retirement. Is this a generational phenomenon, a stereotype, or just a fact of life? The preretiree has more experience, has witnessed more things going awry, and has learned cautiousness and the value of pacing one's efforts. Generally, when folks graduate, they are single and childless. Their priority is to get a good, well-paying job that will help them use what they learned and generate enough money to pay off their student loans. Once these new hires decide to marry and raise children, their priorities shift to make room for these additional responsibilities; work is no longer their primary focus. Similarly, as team members enter their forties and fifties, they start to spend more time thinking about retirement. By this point in their lives, they may have aging parents and school-age children all vying for their time and energy in addition to their career demands. Are these generational differences or just the effects of the cycle of life? Regardless of what you decide, being an effective leader and manager means being able to bring out the best in all of our team members.

BUILDING AND MOTIVATING TEAMS

PRACTICAL WAYS TO BUILD YOUR TEAM NOW (OR ONCE YOU GET ONE)

Remember the adage "People join organizations; they leave bosses." Make it your goal to be the manager who has the reputation of being a talent developer. Strive to be the type of project lead and manager that people want to work for. Following are some suggestions for developing this reputation:

- Communicate and act consistently, so you "walk your talk." Since actions speak louder than words, your actions will enable your team members to understand the factors and criteria you use to define what "good" looks like. For example, if you want everyone to write in a report-worthy way, make sure that all of your correspondence—including your e-mail—satisfies this criterion.
- Recognize each person's achievements on your projects as well as acknowledging the work accomplished on assignments for other project leaders or managers. This gives you the opportunity to set clear performance standards and define work product that is "above and beyond" one's ordinary output.
- Maintain contact and a personalized connection with individual team members as these individuals move from one job to the next. Of course, your willingness to do this will depend on the individual's performance. Assuming that your team members were good performers who have now transferred to other departments, stay in touch with them as a mentor. Keep them in your network.
- View the team member's schedule in total if you are a project lead in a large audit department, not just as it relates to the time this individual is assigned to your projects. Since the goal of all audit departments is to minimize downtime, staff auditors are typically scheduled from one job to another with minimal downtime. Sometimes it takes longer than expected to finish testing and wrap up an assignment. By displaying awareness of the staff's schedule, the staff will feel a sense of loyalty to you and will be inclined to put in the extra effort if needed on your jobs.
- Use experienced staff, where possible, to help plan and run your engagements so these individuals are ready for promotion when that time comes.
- Walk the talk to model how to be flexible, positive, and effective in a changing, deadline-driven environment. When you are in a leadership position, your actions are under constant scrutiny by those reporting to you—especially when unforeseen events occur and wreak havoc with the audit plan. Your colleagues in general and your staff in particular will focus on how you deal with adversity. Insofar as some of your staff may view you as their role model, your behavior epitomizes "acceptable" behavior. Make sure that you are modeling the behaviors you want them to display.
- Help new hires to adjust to the organization, your department, and their job. Since turnover is inevitable, develop a written plan for orienting new hires. Pay particular attention to providing tips and suggestions that will help the new hire adapt to your organization's culture.
- Find opportunities to mentor staff. While some organizations may have a formal mentoring program, the mentor–mentee relationships that occur naturally are among the most genuine and long lasting.

- Base decisions regarding individual performance on observable, fact-based conditions. While you might think that your instincts are exceptionally sharp, when it comes to appraising others' performance, nothing beats having examples and facts on your side. Be prepared to support each observation with specific examples, including copies of work product. Although you think that you have already communicated your feedback in your review comments, the recipient might not have interpreted your comments as a request for sustained behavioral change.
- Help the team members stay excited and energized about the profession, that is, enable each person to correlate the work assignment to the "big picture" within your organization and the profession. I believe that when internal auditing is well done, it is a value-added function within an organization, helping it to achieve its goals while taking intelligent risk. However, when internal audit is poorly done, its results can seem nitpicky.
- Plan stretch assignments—ones that will challenge your team members and help them grow individually and collectively.
- Discuss potential career paths, so that team members can understand the type of growth needed and their possible opportunities; this career path may be outside your department.

Picture it this way: you are the team lead for an audit of your organization's vendor selection and management process. This process has had satisfactory review results for the past three years. Management is tenured, there is low turnover, and the systems and applications are mature and stable. Additionally, the constituent are generally cooperative and meet their deadlines.

Normally, audits take three months, but your team expects this one to be a breeze. In fact, they plan to celebrate at the local bar if it's wrapped up in under two months. You have three more years' experience than the next most senior auditor. Knowing that performance is relative, you're hesitant to rush this audit. Through data collection, you see a potential weakness in the control for returned shipments, potentially costing the company thousands of dollars. You hate to be a killjoy, and your hunch is only that—a hunch. How would an auditor with executive presence handle this situation?

1. Undertake a preliminary, high-level review of multiple variables to present a credible thesis to the team at a lunch meeting you plan with subs from their favorite sandwich shop
2. Silence your doubts and go with the flow

Hopefully, you picked "A". Three important variables go into making this decision: you would feel dishonest quashing your doubts, you made sure to support your stand with facts, and you want to foster camaraderie. At the very least, your hunch is disproven, but hopefully your relationship with your coworkers is strengthened, management noticed your diligence, and you stayed true to your values.

So, you've built your team and you've got them to perform like a world-class orchestra. What will you do when your organization gets a new CEO, your company merges with another, or some other change occurs or opportunity is available?

Following are some tips to help you thrive in a changing environment and encourage positive change in others.

TIPS FOR GETTING TEAMS TO PERFORM

Delegation is defined as investing others with the *responsibility* for accomplishing a task or project and giving them the *authority* to get the job done while retaining *accountability* for its performance.

Delegation can be an effective way to build bench strength among team members. It can also be a way for you to free up time in your schedule by assigning some of your activities to others. You can then use the newly available time to take on other responsibilities. However, delegation involves more than telling someone what to do and using them as an "extra pair of hands." Delegation also involves more than just turning another person loose with just a statement of the desired goal or end result. In order to delegate effectively, you need to understand

- What the desired output or result is, that is, how the result will be used. This information will affect the format of the end product and needs to be communicated to the delegate so he or she understands what he or she is to achieve.
- The delegate's level of familiarity with the subject, problem, or process. The more the delegate possesses subject-matter expertise, the less detailed you will have to be when explaining the action needed to complete the assignment.
- The delegate's work experience and skill level. The more experience the delegate has, the less directive you will need to be. When the delegate has experience, your managerial style can be more collaborative to leverage the delegate's knowledge and ability. Conversely, if the delegate has little experience, consider whether delegating the task is a good idea. If you want to delegate as a means of building bench strength and increasing individual team member skills, then adopt a prescriptive managerial style when making the assignment. Be sure to provide explicit, step-by-step instructions and check in often to assess the delegate's progress (Figure 7.1).

The nature of the task itself—that is, whether it is routine, nonroutine but recurring, or unique—should affect your style when delegating. If the task is routine, the delegate should be able to master it rather quickly because he or she will have sufficient opportunities to practice. Conversely, if the situation is unique, the delegate's mastery may take a very long time, if it happens at all.

HOW TO DELEGATE

1. **Define the Task**: Confirm in your own mind that the task is suitable to be delegated. Do not delegate performance feedback, disciplinary actions, politically sensitive tasks, or confrontations arising from interpersonal conflict.

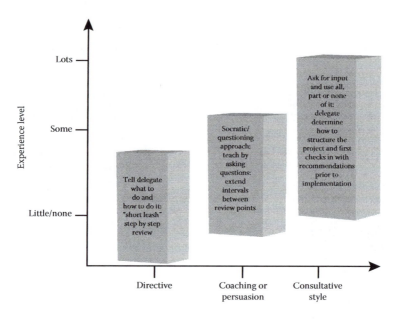

FIGURE 7.1 Delegation and leadership styles. Experience level refers to the delegate's familiarity with the subject as well as the person's work and skill level.

2. **Select the Individual or Team**: What are your reasons for delegating to this person or team? What are they going to get out of it? What are you going to get out of it?
3. **Assess Ability and Training Needs**: Is the other person or people capable of doing the task? Do they understand what needs to be done? If not, you can't delegate.
4. **Explain the Reasons**: You must explain why the job or responsibility is being delegated and why to that person or people. What is its importance and relevance? Where does it fit in the overall scheme of things?
5. **State Required Results**: What must be achieved? How will the task be measured? Clarify understanding by getting feedback from the delegates. Make sure they know your expectations.
6. **Consider Resources Required**: Discuss and agree on the requirements to get the job done. Consider people, location, premises, equipment, money, materials, other related activities, and services.
7. **Agree on Deadlines**: When must the job be finished? Or, if it is an ongoing duty, when are the review dates? When are the reports due? And if the task is complex and has parts or stages, what are the priorities? Confirm the delegates' understanding of the deadlines, priorities, and review criteria. This will give you a sign that the delegates can do the job and reinforces their commitment.

   You and the delegates must agree on the monitoring methods, otherwise the delegates may view this monitoring as interference or lack of trust.

8. **Support and Communicate**: Think about who else needs to know what's going on (e.g., your boss, peers, clients), and inform them. Do not have the delegates inform your peers of their new responsibility. Warn the delegate about any awkward matters of politics or protocol.

9. **Feedback on Results**: Let the delegates know how they are doing and whether they have achieved their aims. If not, you must review with them why things did not go as planned and deal with the problems. You must absorb the consequences of failure and pass on the credit for success.

TIPS FOR MOTIVATING OTHERS WITHOUT MONEY

Yeah, right, you're probably thinking. How can I motivate my team without money? Unless your team members are underpaid for their job level, money will not be a motivator. Similarly, offering to pay someone more to do a job that he or she can't stand will not motivate higher performance levels except in the very short term. Money per se does not motivate. So, if it's not money, what is the motivator?

Three things motivate performance: recognition, achievement, and autonomy. Let's look at each of these motivators.

Recognition means acknowledging an individual's contribution or results. Fundamentally, each of us wants to be noticed and valued for who we are and what we have done. Sometimes the recognition is as simple as saying thank you for a job well done. Depending on the nature of the contribution, an individual may receive an award as part of a group meeting or event. The key to delivering effective recognition is to acknowledge the individual's specific contribution or action. The specificity really matters. General praise isn't that effective. The more specific you can be, including being able to cite examples, the more meaningful your recognition will be. You can use the "tell" and "affect" components of the TACT model covered earlier in this chapter to acknowledge good performers and encourage their development.

Achievement describes the sense of accomplishment individuals experience when they realize that they can do things that they couldn't do before. Essentially, when individuals learn something new (e.g., how to use new software, how to lead a meeting), they are grateful to the teacher. If you position yourself as a coach and teacher, you will become attractive to your team members because you will regularly enable those that report to you to learn new things and experience high levels of achievement. You can use the delegation techniques covered earlier in this chapter as a way to transfer your knowledge to individuals on your team and help them acquire additional capabilities and competencies.

Autonomy means giving others choice and decision-making capability over their schedules and their assignments. In essence, giving others autonomy is the opposite of micromanaging. When you give others autonomy, the goals are predetermined and communicated as are the measures of success. Your team members become responsible for setting their schedules and prioritizing their activities—and in so doing, they become accountable for their performance.

Regardless of how you decide to motivate others' performance, be sure to help people see what influences their rewards, that is, be sure they understand what constitutes good performance. Expectancy theory connects performance effort to rewards through the employees' eyes. In other words, how people perceive the connection between their action and their results influences their behaviors. It is therefore crucial to determine whether and how employees actually see the link and then, through communication, feedback, and appropriate rewards, enable them to see clear, causal correlation. For example, giving employees information about how their department is performing but not showing them how that performance influences their own merit increases is likely to make the information less useful and motivating than it could be.

How to Create Organizational Change

Knowledge and use of the organizational change model provides a framework for making any organizational change a reality (Figure 7.2). Generally, organizational change initiatives require 18–36 months to take hold and 5–7 years to become a seamless part of the organizational culture and practice. Typically, organizational change initiatives require executive management support in the form of a champion. The champion's role is vital to maintaining (and rekindling) momentum as the organizational change is implemented. The champion functions as a cheerleader and arbiter when people need clarification regarding organizational priorities. Given

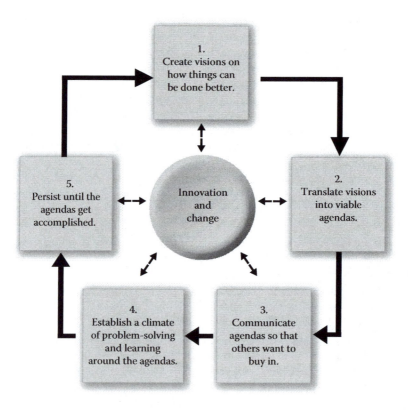

FIGURE 7.2 How to create organizational change.

the time required to make organizational change a part of the culture, choose your initiatives wisely.

Begin your organizational change initiative by creating a vision of how things can be done better or how the team can be transformed. This vision should be of a future state that is several years in the future or it may be a state of aspiration; for example, "the department's members will use leading audit practices." Since leading performance practices evolve, this vision is aspirational and will require a continual commitment to professional development. Make sure that your organizational vision is congruent and compatible with the enterprise's strategic direction and values.

Then translate the vision into viable agendas and projects that can be delegated to teams and individuals to accomplish each year. These projects provide an opportunity to develop project management and planning skills in a nontraditional audit context.

Communicate the agendas so that others want to buy in and contribute to the projects' achievements. The easiest way to do this is to point out the benefits that will accrue to the department as a whole (e.g., more efficient workflow) and the team members as individuals (e.g., skill acquisition).

Perhaps the most challenging step when fomenting organizational change is to establish a problem-solving climate, that is, an environment in which failure and mistakes are viewed as learning opportunities. Those involved in the missteps are not castigated but valued for their experimentation with new behaviors and approaches.

Equally important is to persist until the agendas and projects are completed, which may take 5–7 years. The persistence is critical to prevent team members from viewing the change initiatives as flavors of the month. Given the amount of time it takes to affect organizational change, choose your visions and agendas carefully.

CHAPTER SUMMARY

Understanding how to lead and motivate others is useful information regardless of whether one is a team leader or team member. You don't have to be in a leadership position to use managerial skills. For example, if you have children or are familiar with their behaviors, you have surely seen how at least one child in a group manifests him or herself as the one to follow. If you are perceived as a "go-to" person, then you have demonstrated leadership traits and an ability to influence others. And if you'd like to continue to develop these abilities, try putting the tips in this chapter into practice during the next 30 days.

- Use the attributes of an effective team to guide your team-building efforts.
- Use TACT when delivering feedback.
- Combine leadership and managerial skills when working with a multigenerational team.
- When building a team, remember that people join organizations; they leave bosses.
- When done correctly, delegation builds and reinforces bench strength on your team.
- Money does not motivate; recognition, achievement, and autonomy do.

QUIZ

1. Effective teams (circle all that apply)
   a. Share common values and goals.
   b. Have strong, top-down leadership.
   c. Follow a consistent and logical methodology.
   d. All of the above.
2. The acronym TACT stands for
   a. Tactful Always Considering Ties.
   b. Tell Affect Change Trade-off.
   c. Talk About Change Timely.
3. You should provide feedback as soon as an undesirable behavior occurs to prevent repetition.
   a. True—this avoids repetition of the behavior and quick correction.
   b. False—individuals need time to self-assess and self-correct.
4. It is better to be a leader than a manager.
   a. True—in general, leaders are forward-thinking and managers maintain the status quo.
   b. False—both are needed in a successful organization.
5. When leading a multigenerational team, focus on
   a. Developing skills in those of your generation first.
   b. Teaming different generations to leverage strengths.
   c. Accommodating each team's preferences as much as possible.

The performance planning worksheet

My measurable, time-bound performance goal (i.e., my desired outcome)

Things I want to start doing

Things I want to stop doing

Things I want to continue doing

Observable outcomes and indicators that I've made positive change happen

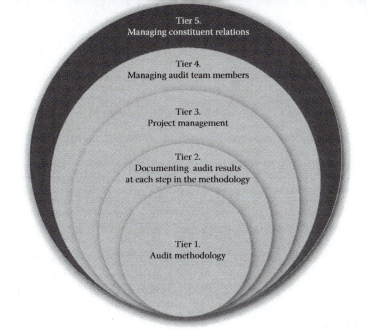

Tier 5.
Managing constituent relations

Tier 4.
Managing audit team members

Tier 3.
Project management

Tier 2.
Documenting audit results
at each step in the methodology

Tier 1.
Audit methodology

8 Techniques for Managing the Constituent Relationship

This chapter covers

- How to create a positive first impression
- Building relationships
- Being influential
- Adapting to others' work approach and behavioral styles
- How to deliver technical information to a nontechnical audience
- How to deliver bad news
- How to soften the blow
- How to deal with stressful and potential conflict situations
- How to differentiate stalls from true objections
- How to overcome objections
- Tips for managing the constituent relationship

If you can't explain it simply, you don't understand it well enough.

Albert Einstein

Until this point, we've focused on the technical aspects of the audit and the skills we've needed as individual auditors tasked with completing the project. In some ways, we've been a bit self-involved, concentrating on what we need to do and how we need to do it. We've focused on our methodology, our documentation, our project, our team, and our logic.

However, by definition, there can be no audit unless there's an constituent. And so, in this final chapter, we focus on the object of our audit, the process owner, who in a way is our *raison d'etre*. Throughout this book, we have used "constituent" to describe this person. As defined in Chapter 1, the constituent is a responsible person(s) in the audited area, for example, the board of directors, executive management, and operating management.

But, if we are to be effective auditors, we need to concentrate on the process owner. We need to make sure that we have established a relationship with this person so that our findings resonate and lead to improvements and change in the organization's risk management culture as well as within the process itself.

Our relationship with the process owner begins the second that person sees or hears us. As superficial as this may seem, the process owner's perception of us is initially based on what we look and sound like. What we have to say and how we conduct ourselves will only support or refute this initial assessment. As unfair as this may sound, it's human nature to judge a book by its cover. That said, there are things that we can do to affect others' initial view of us.

TECHNIQUES FOR ACHIEVING A POSITIVE FIRST IMPRESSION

- Appear approachable by maintaining an "open" posture. This means making sure that your body posture, whether standing or sitting, is inviting and relaxed—that is, hands, arms, and legs uncrossed—no matter how comfortable being contorted like a human pretzel may feel. When taking notes, be sure to leave your notepad in a position that enables the interviewee to easily see what you've written. One key to being approachable is transparency. The person responsible for the process—that is, the object of your audit—should have no doubts or questions concerning you, your approach, or your conclusions. At any point in the audit, your constituent should know exactly where he or she stands. Your goal is to have a no-surprises audit.
- Look people directly in the eye, but do it without staring them down as though you were in a modern day duel. If looking another person in the eye makes you feel uncomfortable, you need to know that this perspective will cause you to lose a valuable source of information.
- Seek out others—don't be a wallflower. By approaching and greeting someone before they can approach and greet you, you are imbued with the aura of confidence—even if you don't feel so confident. Get in the habit of being the first person to greet others. Don't wait for the other person to greet you. By initiating the greeting, you gain control of the social situation.
- Smile: Research indicates that smiling reduces stress and even if this research is wrong, common sense will tell you that smiling makes you

feel happier than frowning. Over the past 10 years, I have performed an unscientific study and the results have been unerringly consistent. When I smile at someone—including a total stranger—the other person smiles back. I've done this experiment so often and in so many different places, for example, my seat on a plane during the boarding process, while walking down a public street, and in hallways in corporate offices. Ultimately, however, the location doesn't matter because the result is always the same: the other people returned the smile. To be fair and balanced in my study, I wanted to see what would happen if I frowned at people or appeared annoyed or irritated. Guess what? One definitely reaps what one sows. Whenever I appeared annoyed or irritated, the other person frequently matched my tone, which would lead to an uncomfortable and sometimes argumentative discussion. I quickly learned that if I wanted a friendly relationship, I needed to be the first one to display friendliness by smiling.

- Shake hands when introducing yourself in person. While it might seem old-fashioned, a handshake creates a human connection—if only for a moment—and it signals your professionalism. The confidence and quality of your handshake speak volumes about you, so make sure that you have a confident grip and are able to make eye contact when you do.
- Say your name and the name of your department slowly, as if you've never said it before. Repeat your first name immediately after saying it for the first time, for example, Anne (pause) Anne Bono. I have a deep voice and, if I'm not careful when introducing myself, others think my name is Andy Taylor instead of Ann Butera.
- Develop a clever way of making people remember your last name; for example, Anne Bono, no relation to the late Sonny Bono. However, be careful; make sure that your word association creates a positive connection.

Talking about What You Do: It is just as it is important to develop a way to make it easy for new acquaintances to remember what you do as it is for them to remember your name. Not what you are, but what you *do*. Simply stating your title or position does not enable someone to have a clear idea of your role as an internal auditor. This is your opportunity to help shape others' perceptions of the value and purpose of internal audit.

I view internal auditors as professionals who:

- Help others manage their business risks and achieve their business objectives without negative surprises
- Act as an organizational mirror, reflecting to management the quality and reality of their risk management practices

In the space provided, jot down an introductory statement that accurately represents what you do and creates a positive image of you in the other persons' minds.

There are no right or wrong answers; this is your opportunity to shape others' perceptions of your work.

```

```

What you say about you attracts or repels others. For example, if I described the role of an internal auditor as the company watchdog, it creates a completely different image and set of expectations. It also sets a tone—and not the best one, I might add—for your working relationship with this person.

Now, you might think, why do I have to say anything at all, other than I work in internal audit? While this sounds like a safe and efficient practice, you are actually missing an opportunity to educate and influence how others perceive the internal audit function. By stating only your job title, you assume that the other person has the correct and complete understanding of your role and your ability to contribute value. You need to accept your responsibility to drive the client relationship and help establish its terms.

Ways to Break the Ice and Fuel the Conversation: Since you are initiating the audit, that is, the contact, it's up to you to break the ice and to establish a relationship.

Start the conversation by posing an open-ended question concerning a neutral topic familiar to the other person, for example, potential weekend plans. Alternatively, you could share something about yourself, for example, the fact that you have a pet or a child, and relate a quick anecdote. Of course, if the other person hates animals or children, never bring up these subjects again. By sharing something about yourself first, you create a psychological indebtedness. Most people respond by sharing something about themselves. This more personal exchange can accelerate the process of getting acquainted.

At each successive encounter, seek to deepen or develop the relationship in a natural way by picking up the icebreaker at the point at which you left off in the prior conversation. For example, if you found out that your client was getting ready to participate in a marathon, be sure to ask about the outcome the next time you get together. Once the ice is broken, keep the conversation moving by asking open-ended questions that require the other person to respond with more than a word or two. Don't just say, "How was your weekend?" Instead you may ask, "What was the best part of your weekend?" And, listen, really listen, to the answer.

Tips for Making Small Talk

Be prepared to discuss several interests or experiences. To identify suitable topics, ask yourself

- What have I read lately that I enjoyed or found thought-provoking?
- What movie, play, or performance tickled my funny bone or captured my imagination?

- What restaurants could I recommend to someone who shares my tastes in food?
- What recordings or concerts have I heard that may be of interest to other music lovers?
- Where have I traveled that exceeded my expectations?
- What new challenges am I setting for myself?
- What are my current hobbies?
- What plans do I have for this weekend or over the next holiday?
- What insights can I share about my business or work that might be interesting?

Do:
- Pick a neutral, yet relevant topic of conversation, for example, the weather, or the commute's traffic conditions.
- Select another topic if the individuals are uninterested in the initial icebreaker.
- Establish eye contact and smile to send receptive signals. (Eye contact for five to ten seconds indicates curiosity and is generally considered friendly. Take care not to stare at another person too intensely because this can make him or her feel uncomfortable.)
- Be the first to introduce yourself and ask an easy, open-ended question.
- Listen carefully for facts, feelings, and key words.
- Highlight mutual interests.
- Act sincerely.

While you can choose any topic you like as an icebreaker, your goal is to get a conversation started—and more fundamentally, to get the other person to appreciate you as an individual, not just as the internal auditor. Consequently, you should avoid the following subjects:

- Stories of questionable taste—and if you are not sure what constitutes "questionable," it probably is, so don't say it.
- Gossip—if you talk about a coworker who is not present (especially in an unflattering way), the other people will assume that you will talk about them in their absence.
- Personal misfortunes—while these sad situations are on your mind, you do not want to weigh down a budding relationship by sharing this information.
- How much things cost—your goal is to create a relationship by identifying and focusing on areas of commonality. Discussing the cost of things, that is, quoting a dollar amount and then characterizing it as cheap or expensive, without knowing the other person's perspective is risky.
- Controversial subjects, such as politics, if you don't know where everyone stands on them. You could be at odds with their views. Your goal is to start your relationship in the state of mutual agreement.

- Bad news like terrorism, war, pestilence, and famine. These topics can cast a pall over the discussion, especially if it's one where you're planning to deliver bad news. However, they are worthwhile topics to discuss and evaluate, and you may be able to do so if you have a relationship with the person outside of work. Some other topics in this category include divorce, death, layoffs, gloomy economic predictions, and the Red Sox versus Yankees (at least in Boston or New York!).
- Health (yours or theirs) unless both of you are hale and hearty. Your icebreaker should create a fluid and upbeat exchange. It's not the time for detailed conversations about medical test procedures and results.

The following are some tips for fueling a conversation:

- Be the first to say hello.
- Be able to succinctly tell others what you do.
- Make an extra effort to remember and use people's names.
- Be aware of open and closed body language. If your arms and legs are crossed or your hands are folded, you could unwittingly be signally that you are closed to what the other person has to stay. Become aware of where your body parts are during the conversation.
- Seek the opinions of others. Make sure to ask open-ended questions to get the conversation started. Most people love to tell you what they think; all you need to do is ask.
- Look for signs of boredom or restlessness from your listener and take this as the signal to get down to business.

So now imagine that you have successfully broken the ice and set a comfortable tone for the discussion or meeting you are about to have.

The opening minutes of your presentation are very important. During this time, you have to grab and focus the person or group's attention and kick off the meeting or discussion. The following are several things you can do to start with strength:

1. Walk confidently to the front of the room; if you will be speaking while standing, place your feet shoulder-width apart and smile. (If seated, make sure that you are sitting upright in your chair and both feet are on the floor.)
2. Speak in a conversational tone with confidence, enthusiasm, and energy.
3. Rehearse the first 1–2 min of your message, until you don't have to think about it.
4. Know and describe the purpose and objectives of the meeting and your message without checking your notes.
5. Allow your arms and legs to move naturally when standing. Keep a relaxed, upbeat facial expression. Refrain from resting hands on face (if seated) and hips (if standing). Keep your hands out of your pockets. Avoid the "fig leaf" position when standing, that is, hands crossed in front of your body.

Other Tips for Breaking the Ice and Creating a Positive Atmosphere are

- Greet the meeting attendees as they enter the room, before the session starts. If you are convening a large conference call, that is, one with more than four participants, introduce yourself and ask everyone to mute the phone lines until the meeting starts.
- Ask informal questions before the meeting starts to discover information about the audience's background, interests, hobbies, and so on, or to expand on an existing relationship.
- Use humor as a means of putting the audience at ease. However, if you are not good at telling jokes, don't even try.

INFLUENCING TECHNIQUES

Once you have established the relationship and gained trust with others, you are in a position to influence. By definition, influence is the ability to get others to act on your suggestions without pulling rank. Influential people are able to garner support for their ideas. They understand that being persuasive requires more than technical expertise and simply having facts to support a perspective. They are able to communicate their message in as many ways as necessary to appeal to the diversity of their audience. Persuasive people leverage their relationships with others and the information they possess to get others to act on corrective action plans and implement suggestions for increased efficiency.

Influential people do not require a formal title and those with formal titles are not necessarily persuasive within their organizations. You can identify influencers by noticing who people gravitate toward for information during a project or periods of organizational change and unrest.

Also, persuasive people project values that are meaningful to those they influence. They identify commonalities with others to establish and build relationships. Then, they leverage these relationships to persuade others and gain agreements. At the top echelons in an organization, the ability to influence equates to the ability to get things done.

Are you an influencer? Would you like to be more so? The following five techniques will increase your ability to influence others' behavior:

1. Identify and expand areas of commonality when interacting with others. Take the time to break the ice with the folks with whom you work. During this time, pay attention to their answers to common conversational questions like "How was your weekend?" These answers can provide insight into their hobbies and priorities. To the extent that you genuinely share interests in these areas, your ties to these individuals will expand and deepen over time.
2. Pace before leading others. Take the time to understand the other person's preferred communication style and match it. For example, if the other person is a slow, methodical, and precise communicator, your use of overblown or imprecise words, for example, always, never, and very, will trigger

suspicion. Likewise, speaking very quickly to this same individual will engender distrust. Instead, choose your words with care and slow down your rate of speech. Establish this rapport before diving into an explanation of your ideas.

3. Watch your language. Use common, everyday terms to explain technical concepts. Avoid audit jargon, for example, inherent risk, residual risk, and key control. Be prepared to express the same message in several ways until the other person understands what you are saying.

4. Adapt your communication style. If you are dealing with an analytical person, present your position in a coherent, sequenced manner. If you are dealing with a goal-oriented person, explain how your ideas will enable this person to achieve his or her goals effectively or efficiently. If you are dealing with a people pleaser, provide examples of precedents that illustrate how other departments or teams have successfully implemented your suggestions. If you are dealing with a high-energy strategist, keep your messages focused, concise, and simple by limiting the amount of detail you provide.

5. Pull more and push less. Instead of making statements and telling people what to do, use questioning to engage the others. Most people believe "their own baloney," that is, if they say it, they own it. This means that if they identify a gap in their process or a breakdown in their process's controls, they believe these conditions exist. When you deliver the same message and tell them they have a process gap, their typical reactions will be defensive and resistant. Consequently, take the time to devise a series of open-ended questions that will lead the people you want to persuade to arrive at the point you want to make—without you having to explicitly tell them what you would like them to do. When you use this questioning approach, you enable the respondents to think through their responses and use their own words to describe the condition you want addressed. While this indirect approach may appear to be time consuming, it achieves results that are long lasting.

While these techniques may seem both simple and simplistic, they require a great deal of self-control and practice before they become second nature. However, if you make it a habit to apply the five techniques listed above, you will become an influential team member within your organization, gaining the trust of your peers and those you audit.

STRATEGIES FOR INCREASED INFLUENCING ABILITY

Follow these steps to increase your ability to influence others:

- Understand and empathize with others' needs.
- Be instrumental in defining the roles of others and creating frameworks for accomplishing results.

- Develop a track record for arriving at fair and equitable "win-win" solutions.
- Develop a track record for getting results without bloodshed
- Foster climates of open information exchange among all parties.
- Expand your vocabulary so you have an array of expressions and words to use.
- Practice verbal agility to become proficient in it
- Demonstrate values and behaviors that are congruent with the corporate culture.

Influencing skills are the antithesis of critical thinking skills.

- Influencing is the act of persuading others to see a certain point of view or change an existing attitude. To do this, the influencer communicates messages using other-directed language and words that will resonate with the other party. The influencer listens to the other person and adjusts the message as needed to reach agreement.
- Critical thinking is an inward process that helps you deduce information from verbal and non verbal communication. It is an internal, analytical process.

Influencers	Critical Thinkers
Use general terms, for example, most, usually, typically, everyone, no one	Use specific terms and cite specific amounts
Emphasize data related to shared values and commonalities	Focus on all data objectively and equally

DEALING WITH DIFFERENT BEHAVIORAL STYLES

While each person is an individual, people have preferred communication styles. Some are more introverted than extroverted. Others are more results-oriented than people-oriented. Some speak and move more rapidly than others. Some prefer to approach situations from a relationship-oriented way, and others through a task-oriented way. To be clear, there isn't one way of being that fits internal auditors better—rather, we need to leverage the best in each person by adapting our own style. To do this, we need to understand our style and recognize others' styles. Since we can't get into people's heads, we have to pick up clues by observing their behaviors and their proclivities when approaching a task.

Behavioral Styles: Throughout my career, I have assessed and profiled thousands of auditors' behavior ranging from staff auditors to chief audit executives. This assessment process enables individuals to categorize their own behavioral preferences and recognize others' preferred styles. The behavioral styles are summarized in this section. As you review this information, keep in mind that these are profiles. Also, some people display characteristics of more than one behavioral style depending on the situation, for example, personal versus professional relationships. But generally, I've found that most people (auditors and their constituents alike) fall into one of the following four categories:

- People-pleasers
- Analyticals
- Bottom-liners
- Vocalizers

People-pleasers are warm, friendly, and helpful. They are concerned about precedents, fairness, and relationships in general, especially those between themselves and others. Before making a decision they may look for help from others to build consensus and create agreement. They are collegial and diplomatic, striving to please the majority, if not the whole.

Typical Traits:

- Desires to be liked and accepted
- Wants you to be pleasant and friendly
- In a meeting is harmonizing and helpful
- Gets irritated by others' impatience and pushiness
- Pace and timing of interactions and conversations is comfortable and slow
- Looks for attention
- Time style focuses on then and when, that is, the past
- Decisions are considered

How to Win Them Over:

- Impress them by your friendliness
- Ask questions that are non threatening
- Support their feelings
- Demonstrate your team spirit and warmth
- Make benefits personal
- Show commitment by working with them
- Be impressed by their loyalty
- Best close to use is assumptive

Analyticals are likely to speak in steady, even-keeled, measured tones and display interest in the details of things. They want arguments proved logically, with figures to back up claims, before deciding.

Typical Traits:

- Desires to be accurate
- Wants you to be precise and accurate
- Channels discussions and clarifies comments during meetings
- Gets irritated by surprises and unpredictability
- Pace and timing of interactions and conversations is systematic and slow
- Looks for accuracy and detail

- Time style focuses on scheduling
- Decisions are deliberate

How to Win Them Over:

- Impress them with your thoroughness
- Ask questions that are detailed
- Support their thoughts and approach
- Demonstrate your detailed knowledge
- Make benefits provable
- Show commitment by being systematic
- Be impressed by their accuracy and attention to detail
- Best close is a detailed summary

Bottom-liners are likely to be direct, succinct, and abrupt. They want to get to the point without any "beating around the bush." If they see the benefit, they will make immediate decisions. Bottom-liners are typically in high-managerial positions or even run their own business.

Typical Traits:

- Desires to be in charge
- Wants you to be to the point
- Initiates and directs during meetings
- Gets irritated by others' indecision and slowness
- Pace and timing of interactions and conversations is decisive and fast
- Looks for productivity, results, and achievement
- Time style focuses on NOW!
- Decisions are closed and final

How to Win Them Over:

- Impress them by getting to the point
- Ask questions that are relevant
- Support their actions
- Demonstrate your experience
- Make benefits tangible and concrete
- Show commitment by getting things done
- Be impressed by their strengths
- Best close is direct

Lastly, Vocalizers are effusive and talkative, with a constant flow of thoughts and ideas. It can be hard to keep up with them, or get a word in edgeways! They tend to talk to think, so be sure to use a lot of restatements to help them clarify their ideas before you take any action. They have many ideas, which make them great "intenders" and procrastinators, so you have to pin them down to a commitment. Vocalizers tend to be heuristic thinkers, which can cause others to view them as disorganized or flighty. However, beware of writing Vocalizers off—they secured a position in your organization for a reason, and it may be that you have to spend more time than expected to adjust your style, especially if you are an Analytical or a Bottom-liner.

Typical Traits:

- Desires to be admired
- Wants you to be insightful
- Explores and questions while in meetings
- Gets irritated by routines and the mundane
- Pace and timing of interactions and conversations is spontaneous and fast
- Looks for recognition
- Time style focuses on delayed decision-making and procrastinating
- Decisions are spontaneous

How to Win Them Over:

- Impress them by your flexibility
- Be impressed by their flexibility
- Ask questions that are far-reaching in their implications
- Support their ideas
- Demonstrate your originality and creativity
- Make benefits innovative and global
- Show commitment by providing fresh ideas
- Best close is direct

Diagnosing Others' Styles

To put this information to use, you need to figure out your preferred communication style before you focus on others. How and what you communicate can affect others' styles. For example, if you are too strong a Bottom-liner, you can unwittingly cause others to shut down. Once you're aware of your own behavior, focus on what and how others are communicating.

By honing your observation skills, paying attention to several behavioral indicators, and practicing a bit, you'll be able to determine other people's preferred communication styles.

The key behavioral indicators to observe are:

- Speed in speech
- Speed in gestures
- Speed in response time
- Degree of vocal variety
- Degree of facial expression
- Inherent loudness

Let me illustrate with two examples: one, who we will call Olga, is a Vocalizer. Her delivery is rapid-fire, she answers questions before they are fully articulated, and her team knows to "just do it." Even her walking rate says a lot—big, purposeful strides that cause people to step out of the way when she appears. She has honed her personality so that her delivery doesn't sting—her team knows not to take it personally, but to do their best in the most efficient way possible. There are tight deadlines to meet!

Peter is an analytical. There are less inflections in his speech, and he gives the impression that every word is carefully chosen. He is slower to respond to an inquiry made in person than via e-mail. (This is not a question of intelligence, rather a sign of how he processes information.) Even taking his lunch order is a process, since he is a conscientious eater. His clients and colleagues know that even though he is reserved, any information that comes from him is well thought out. They know to probe more if he says something that doesn't initially make sense because his line of reasoning is linear, and understandable—even if they don't agree with it.

There is merit to observing body language and tuning in to speech patterns because they can give you insights into the preferred communication style of those with whom you are dealing, clue you into the group dynamic, and enable you to identify who may need more coaxing.

DELIVERING TECHNICAL INFORMATION TO NONTECHNICAL PEOPLE

FIVE TIPS FOR MAKING THE ABSTRACT CONCRETE

While it is helpful when influencing others to identify and adapt to the other person's communication style, it is also important to understand their subject-matter expertise and familiarity with technical concepts and jargon. The following tips will help you communicate technical messages to others who do not have technical expertise.

Tip #1

Ask yourself

- What ideas can I visually illustrate or demonstrate, instead of describing them only in words?
- What examples can I provide to illustrate a process, problem, or result, as opposed to describing it only in words?
- What abstract ideas can be posed to my audience as a question? Sometimes questions force people to think in practical terms.

Tip #2

Eliminate as much technical terminology as possible from your discussion. Nontechnical audiences will inevitably have a difficult time with words, acronyms, and phrases they don't normally use. If it is impossible to eliminate such terms, be sure to explain them at the outset. Try to transform technical jargon into street talk. Street talk includes what you hear on television, advertising, radio, and what you read in the newspaper. It also includes buzzwords and catch phrases. Try to:

- Make use of simple analogies and metaphors to describe complex ideas.

- Make use of humor or jokes to convey the description, process, or significance of complex ideas.
- Draw references from everyday life: commercials, popular products, cultural trends, current event stories, TV shows, movies, and songs.
- Quote famous people—use the slogans of politicians, corporate advertisers, or entertainers, if pertinent.

Tip #3

Since the average attention span for most adults is 10 min, aim to keep all technical topics short and to the point by using the following techniques:

- Try to state your point in one sentence. Think ahead of time and try to reduce your thoughts to their common denominators.
- Consider making three points per segment. We are all familiar with memorizing things in a three-point process.
- Imagine that you are creating a sound bite (interview clip) on a documentary or news broadcast. Try to limit your point in a certain allotment of minutes or seconds (e.g., give yourself a one-minute segment).
- Try to extract the essence of what you are trying to say and eliminate supporting data, unnecessary commentary, or tangential information.

Tip #4

If you are having difficulty focusing your ideas, ask yourself: What's my goal in making this presentation? Consider the following:

- To inform: To pass along new information, statistics, data, or opinions to a group of people who may or may not have a direct stake in your presentation.
- To educate: To instruct others by providing them with background knowledge, theories, techniques, and how-to information on a particular subject matter.
- To persuade: To convince a group of people to support your idea, or to get them to change positions on a particular issue or recognize the risk.
- To solicit some form of action: To convince a group of people to do something, such as take corrective action.

Tip #5

It's easy to lose your audience if your presentation has no logical sequence. Consider the following ways to sequence information:

- Chronological order: In what sequence does the process occur? It helps to use units of time (i.e., minutes, hours, dates, years, etc.) to assist you.
- Problem > Cause > Solution: Present a problem; state its cause; and then describe its solution.
- Acronyms and memory devices: Present a series of letters that order your points. (This should not be overused.)

- Comparison and contrast: This is useful for presenting the advantages and disadvantages of something.
- Deductive reasoning: This sequence is based on if-then statements. State your premise, add the facts, and then make your conclusion.
- Inductive reasoning: Present a series of facts or points of information, and then make your conclusion based on what you have presented.

When presenting your observations and results:

- Focus on the facts, not how you feel about them. Anticipate objections and plan how you will counter them by using your knowledge of the area.
- Be prepared to receive at least seven objections before reaching agreement.
- Define the benefits, that is, why management should take the corrective action, by thinking through the impact on business goal achievement if the situation continues unaddressed.
- Answer the question: "So what—why should anyone care about taking action?" Consider the effect on the company's reputation and customer service as well as whether the company could incur brand damage, sanctions, fines, or penalties.
- Present your results within the context of the "big picture." Correlate your results to the business challenges and priorities.
- Make sure you have "sold" the risk before trying to get agreement on corrective actions. Be able to express the same concept in several ways until the other person understands your message.
- Obtain the other person's agreement to take remedial action. Reaching agreement means getting a "yes" or a "no" from the other person. "Maybe" is not an acceptable response and does not mean that the other person agrees.
- Pay attention to the other person's reaction and respond to it.
- Be prepared to supply the detail—if asked. Have copies of pertinent workpapers.

TECHNIQUES FOR DELIVERING BAD NEWS

The delivery of bad news calls for tact, diplomacy, and concern for the well-being of both you and the other person. To incorporate these qualities into a conversation in which you must deliver bad news, consider taking the following actions:

1. Pick a time and place when you can be free from distraction or interruption.
2. Get right to the point. Announce up front that you have some unpleasant, unfortunate, disappointing, or disturbing news. The right words? Simple: "I have some unpleasant news."
3. Use "softeners" to open. For example: "I'm sorry to have to tell you..." or "I'm afraid that..."
4. If the news is coming as a shock to the other person, be prepared for their emotional reaction. Let them vent, if they seem to need to. Do not try to get them to "calm down" or "be reasonable."

5. If you are concerned about their reacting with violence, make sure you have provided for your own safety and security. Either have a witness present, or alert security in advance.
6. If appropriate, once the shock has abated, offer the person resources they can pursue.
7. Forgive yourself for being the bearer of bad news. You are not causing their distress—the news is.

TIPS TO SOFTEN THE BLOW

While it's never pretty, there are things you can do to soften the blow of bad news. The next time you have to present less-than-favorable information, keep the following tips in mind.

- Tailor your message appropriately and eliminate adjectives and adverbs. Just convey the facts without embellishment.
- Tailor your presentation appropriately. You wouldn't wear a Hawaiian print shirt to a funeral, so don't use bright colors, cartoons, sound-effects, or zany fonts if your PowerPoint presentation contains a series of grim statistics. Stick to a simple background color (or use a standard corporate template) and sans-serif font. Save the pulsing and undulating transitions and animation effects for a more upbeat presentation.
- Don't invite extra spectators. When you schedule a bad-news meeting, it's particularly important to invite only those people necessary to the discussion. When you discuss control breakdowns within the billing group, call a meeting with pertinent managers and VPs. Give them the facts, then leave it up to individual managers to disseminate the information to their teams.
- Don't be overly dramatic. Okay, there are some test exceptions and control breakdowns. While this is disappointing to the area manager, it's not the end of the world. Report the facts, get the manager's agreement on them, and don't exaggerate the impact of your findings.
- Include a positive spin. Bad news is always easier to swallow if it's delivered with a positive spin. For example, if you must report that the results of your audit are less than favorable, you'll also want to include some positive news, for example, that local management has already begun to take action... if this is true.
- Don't sugarcoat it. On the other hand, be careful not to put too much of a positive spin on the information. You have an obligation to share the facts—even if they're alarming or upsetting to others within the organization. After all, it's business. Numbers fall, campaigns fail, employees don't work out, and the economy slumps—people cope with bad news every day. Be forthright, objective, and optimistic—it's the best way to deliver bad news.

Once you've conveyed your message, be prepared for pushback and objections because this is the natural human response to bad news. If after hearing the bad news, the other person says nothing or agrees without argument, excuse, or comment, one of two things has happened:

1. The other person is completely in denial, triggered by the shock of hearing the bad news. This person has temporarily shut down and needs time to process your message. Once this person has had time to comprehend your message, expect pushback and objections or at least an attempt to minimize the impact of your findings.
2. The other person is not an accountable party, that is, not the process or risk owner. You are not getting any reaction because this person has no skin in the game and will not experience any fallout as a result of your bad news. When this occurs, you need to figure out to whom you need to escalate this news. Sometimes, the easiest way to do this is simply to ask.

Since objections are a natural reaction to the receipt of bad news, you should be prepared to deal with them.

TECHNIQUES FOR OVERCOMING OBJECTIONS

It's helpful to view objections objectively. They are reasons or arguments presented in opposition to a point or proposal. Objections occur naturally whenever you are trying to influence or persuade someone else. There are two ways to handle objections:

1. *Preventively*: This means that during your planning and analysis you anticipate the other person's objections and preempt them. Preempting objections means that you state the objection before the other person can express it and you state how it can be overcome.
2. *Prescriptively*: This means that you counter the objection after the other person has mentioned it. The prescriptive approach can sometimes create the impression that you must defend your position. This impression can undermine the potency of your position—especially if you are unprepared to deal with the objection.

In general, it is better to prevent objections than to have to deal with them once the other person has brought them up. Being proactive enables you to control the objection's positioning.

FUNDAMENTAL RULES FOR HANDLING OBJECTIONS

When dealing with objections you should:

- Never tell the other person, "What you need to understand is…."
- Never allow the other person to put you on the defensive. Defensiveness is a state of mind. You can control your states of mind.
- Never let an objection go. You might have to temporarily table the conversation, but do not allow the other person to think that you have ceded the point.

- Always listen—really listen—to the objection. Be sure you understand the trigger for the other person's reaction.
- Always acknowledge the objection by restating what the other person has said and then state your point of view. This way you have given them some assurance that you heard what they said before you offer your perspective.
- Always be prepared to prove your position with testimonials, reference sources, and corroborative documentation, especially if you are communicating with an Analytical.
- Always remember that objections are a natural part of making change happen. Anticipate and welcome them.

Four Questions to Determine the Source of Objections: When dealing with objections, it is very important to differentiate between excuses or stalls and actual barriers that need to be overcome in order to resolve the situation. The best way to do this is to ask:

- "What other information will you need before making a decision?"
- "What other considerations will you want to think about before going ahead?"
- "What other concerns do you have that we need to talk about?"
- "What other questions do you have that need to be answered before you feel comfortable making a decision?"

How to Overcome Objections Once They Are Raised: When the other person objects to your message and pushes back, consider taking the following actions:

Listen carefully to determine the source and cause of the objection. Tone down your approach immediately. The more you care about an issue, the less likely you are to be on your best behavior. Believe that the other person might have something to say. Probe to clarify your understanding of the message, for example, "Why would this recommendation make it hard for you to do business?" Restate your understanding of the objection so that you prevent misunderstandings and assure the audit customer that you have been listening.

Address the objection by using one or more of the following approaches:

- Correcting erroneous audit customer beliefs or assumptions
- Clarifying the nature and extent of required changes
- Reiterating the business risk using different expressions, examples, or war stories
- Quantifying or qualifying how the benefit of accepting your position will offset the audit customer's business risks

Ask the other person whether you have overcome the objection. If the response is no, ask what information would overcome the objection. Then seek to provide that information. Successful sales people will urge you to be prepared to handle seven— nine objections as a natural part of the process. Most people don't realize this and are very surprised when an objection they thought was resolved and closed down is reintroduced for discussion.

EXERCISE 8.1: OVERCOMING CONSTITUENT OBJECTIONS TO ISSUES

Instructions: Describe typical objections you encounter when discussing issues. After you have listed these barriers, use techniques discussed in this chapter to develop ways to overcome barriers.

The Other Person's Objection	What My Response Will Be
Too Little Money: "There's no money in the budget to fund what you are suggesting."	
Too Little Staff: "The Staff is over- committed now."	
Busy Staff Schedule: "These are our peak production periods."	
Scapegoating: "It's not up to us; it's the responsibility of <name of another department, company, or person>."	
The Residual Risk is Misunderstood: "We are willing to assume that risk."	
Dilettante: "Just because it's in a book, for example, an industry standard, it doesn't make it right."	

TIPS FOR MANAGING THE CONSTITUENT RELATIONSHIP EFFICIENTLY

Consider using some of the following techniques to establish or expand your relationship with the people you audit.

Evaluate the situation to determine:

- The decision-maker, decision-influencers, and decision-making process
- The impact of the risk or issue on the business objectives and needs
- The time frames within which action is required and will happen
- The potential budget implications

Think about and satisfy the behavioral style of your audience, particularly the decision-makers and decision-influencers.

Create a running list of questions and then contact the constituent every few days to address them. During the interview, present yourself in a professional manner and demonstrate that you are familiar with the business or the area being audited. Pose your questions in a logical order. Compliment the constituent and encourage open discussion by using open-ended questions, maintaining eye contact, and looking interested.

To bridge language barriers, remember that "a picture is worth a thousand words." Use diagrams to exchange information. Develop and plan how you will communicate a verbal framework that lets the other person know the discussion's purpose, desired outcomes, and the number of issues to be covered. At the start of any communication, the other person must know what he or she will be expected to do.

Summarize the results of each prior conversation or meeting before proceeding to cover "new" material.

Confirm agreement throughout the course of the audit. Don't wait for the exit meeting to begin solidifying customer agreement concerning risks, controls, and process gaps. Get commitment dates for corrective action plans as soon as possible. Don't oversell. Stop presenting once mutual agreement is reached.

Don't wait for the final meeting or exit conference to present issues. Make sure the right people are at the table early. As preliminary findings emerge and you begin to draw initial conclusions, informally begin to discuss your results with decision-makers and decision-influencers. Watch their reactions and make appropriate adjustments to your word choice.

When language is a major barrier, communicate using simple terms; don't get flowery or use jargon. Be aware that others may use terminology that is different than yours. Alter your vocabulary to include words used by them. Experiment with different ways of expressing the same thought to obtain information and understand what they are saying.

Avoid using audit terminology, for example, "properly accounted for" and "risk issue." Set realistic expectations concerning the scope and depth of information to be covered during an interview. Keep in mind that an interviewee will not know everything about the area or topic under discussion. When you receive tentative or noncommittal responses, do not jump to conclusions that a control weakness exists. Verify open issues by speaking with others.

Maintain eye contact during interviews. Do not allow your notes or note-taking to become an obstruction to the exchange of information. Listen actively. Demonstrate "professional skepticism." This means that you are focused on making sure that there is evidence, an audit trail, or a history file to support the information you collect. To this end, you ask, "How do you know?" to identify the evidence that controls are functioning.

Obtain supporting material (evidence) to substantiate the interviewees' information. Verify and confirm your understanding of the information each interviewee gives you by using restatement and paraphrasing. Don't walk away without understanding the information you have received. At the end of the interview, summarize the information. Thank the interviewee and leave the door open for further questions. Indicate to interviewees that after you have summarized and documented the workflow, you will send them a copy so that they can confirm the accuracy of the information.

Review the relevancy, accuracy, and completeness of your notes with the interviewee before you leave the meeting. Do not promise a copy of the meetings minutes or other information if you don't really plan to send it or if you are not authorized to do so.

CHAPTER WRAP-UP

People management is crucial at all stages of an audit. By building a healthy working relationship from the start, you set the foundation to be influential and to communicate effectively at each stage of the audit. By paying attention to body language and speech patterns, you will be able to diagnose the other person's preferred communication style and adapt to it. Being able to adapt to other's communication style will help you deliver bad news without creating bad feelings, overcome objections, and prevent conflict.

CHAPTER SUMMARY

- First impressions do count! It's in your hands to create a positive one.
- Listen, really listen.
- Use fact, reaction, analysis, and application questions to facilitate discussion.
- Establish the relationship and gain trust to boost your ability to influence others.
- Adjust your communication style to other people's behavioral styles because it will make it easier for them to understand your message.
- Extract the essence of what you are trying to say and eliminate unnecessary commentary and tangential information when communicating your ideas to others.
- Eliminate jargon and use common analogies when presenting technical information to nontechnical people.
- Reach agreement with business management concerning the root cause of the objection or conflict before responding with a potential solution.

QUIZ

1. It's right to "judge a book by its cover."
   a. Never, we must give people time to display themselves.
   b. It doesn't matter if it's right—it happens, and we need to make sure we reached the right conclusion.
   c. Absolutely—what you see is what you get.
2. Since you are the auditor, it is your responsibility to establish the relationship.
   a. True—I am initiating the project, and I can set the tone for the next few months.
   b. False—audit requires independence, not the ability to influence.
3. It's important to talk to the people about something other than the audit.
   a. True—this leads to mutual trust and a camaraderie based on real feelings.
   b. False—work is work—everything should stay in its place to avoid blurry boundaries.
4. The most important aspect of facilitating is
   a. Setting the foundation by creating a proper agenda.
   b. Focusing on how to bring together of diverse backgrounds, interests, and capabilities.
   c. Making sure you are completely prepared.
   d. None—they are all important.
5. Your communication approach should stay the same at all points of the audit.

a. True—the client and your team needs to feel that you are consistent.

b. False—different stages of the audit may require your communication style to change.

The performance planning worksheet

My measurable, time-bound performance goal (i.e., my desired outcome)

Things I want to start doing

Things I want to stop doing

Things I want to continue doing

Observable outcomes and indicators that I've made positive change happen

9 Chapter Quiz Answers

CHAPTER 1: HOW TO GET THE MOST FROM THIS BOOK

1. A competency is mostly based on
   a. Innate ability.
   b. Skills acquired through training.
   c. *Skills developed from experience.*
2. If you want sustained results, you should
   a. Work on all your desired changes at once.
   b. *Work on changing one or two behaviors at a time.*
   c. Hire a professional coach.
3. The three types of change are
   a. *Macro, micro, and organizational.*
   b. Situational, focused, and widespread.
   c. Personal, business, and foreign.
4. To achieve sustained change, you must target the right behaviors.
   a. *True*
   b. False
5. A strength can become a weakness when
   a. It is not used enough.
   b. *It is overused.*
   c. It clashes with others' abilities.

CHAPTER 2: TECHNIQUES FOR PLANNING USEFUL AUDITS

1. Audit planning
   a. Is the act of documenting all the process steps.
   b. *Comprises the activities needed to acquire an understanding of the area under review, set the audit's objectives and scope, determine the resources required to complete the project, establish the timelines and deliverable due dates, and officially kick off the audit with the area's management.*
   c. Is a phase only relevant to audit leads.
2. The steps in the Critical Linkage™ are
   a. Interview, document, test, corrective action.
   b. *Objectives, risks, controls, corrective action.*
   c. Plan, discuss, test, report.
3. To gain perspective of the area you need to audit, you must focus on
   a. The business area overall.
   b. The external environment.
   c. The internal environment.
   d. *All of the above.*

4. T/F The data you need to plan an audit will only come from interviewing process owners, for example, senior local management and other informed parties.
   a. True
   b. *False—secondary sources like prior audit reports, the business's strategic plan, results of self-assessments, management reports and any other published data will provide key information and a more robust entity profile.*
5. The funnel approach is
   a. Asking "why" until you get to the essence.
   b. Asking "who, what, when, where, why."
   c. *Asking open-ended questions, making restatements, asking closed-ended questions, and using "what if" constructions.*

CHAPTER 3: TECHNIQUES FOR DETAILED RISK AND CONTROL ASSESSMENT

1. A synonym for "risk event" is
   a. *Risk.*
   b. Cause.
   c. Impact.
2. Step 1 in identifying risk according to the Risk Mental Model is to
   a. Ask, How does management control the risk?
   b. *Ask, What could go wrong to threaten the business, function, or process objective?*
   c. Ask, What is the process trying to achieve?
3. Understanding the risk's root cause
   a. *Is important because it will make it easier to identify the types of controls that are necessary to prevent or detect and correct the risk should it occur.*
   b. Is important because it will clarify the process objectives.
   c. Is unimportant because there are many ways to control risks.
4. When assessing risks, you should consider the effect of controls that are in place.
   a. True
   b. *False; focus on performing a "gross" risk assessment.*
5. As a general principle, the higher the risk, the more critical it is to have control activities in place and functioning.
   a. *True*
   b. False
6. The term *controls* encompasses:
   a. Control environmental elements.
   b. Control activities.
   c. Monitoring.
   d. A and B.
   e. *All of the above.*

CHAPTER 4: TESTING AND SAMPLING TECHNIQUES

1. When determining a testing approach, you can
   a. Test the entire population.
   b. Focus on high-risk areas.
   c. Use sampling.
   d. *All of the above.*
2. The first step when developing control tests for operating effectiveness is
   a. Selecting the controls to be tested.
   b. Gaining an understanding of the control design.
   c. *Determining test objective.*
   d. None of the above.
3. Sampling is a mathematically derived testing procedure that will enable you to quantify the sampling risk
   a. True
   b. *False*
4. Substantive testing is never used.
   a. True
   b. *False*
5. Regardless of how many times the control is performed and the population size, it is best to pick 30–40 items or the most possible.
   a. True
   b. *False*
6. An isolated incident is
   a. The same as an exception.
   b. *An aberration that is infrequent, a one-off, or insignificant.*
   c. An aberration or deviation that is systemic, widespread, or material.

CHAPTER 5: DOCUMENTATION AND ISSUE DEVELOPMENT

1. Summary writing is
   a. *The expression of opinions and ideas based on review and analysis.*
   b. Documenting the information obtained.
   c. Filling out all fields in a template.
2. I should pay the most attention to writing
   a. When all issues have been identified and the testing documentation is compiled.
   b. *Starting with the first workpaper I draft.*
   c. When drafting the final audit report.
3. An issue is
   a. A description of what you found.
   b. *A description of control weakness identifying the root cause of the problem.*
   c. A description of the effect or consequence.
4. Well-written audit issues must contain
   a. Condition.

      b. Cause.
      c. Criteria.
      d. Consequence.
      e. Corrective actions.
      f. All of the above.

5. When identifying issues, avoid discussing them with the area or process's management
      a. True—discussing may slant issue identification.
      b. *False—this discussion will help determine the cause of exceptions and audit issues.*

6. An effectively written audit report should
      a. Summarize the contents of control evaluations.
      b. Describe all test results.
      c. *Trigger actions that resolve the issues raised.*

CHAPTER 6: CORE COMPETENCIES YOU NEED AS AN AUDITOR

1. Executive presence is about the image we portray.
      a. True.
      b. *False—it also includes innate qualities like knowledge and intelligence.*

2. Critical thinking is not crucial if you are not leading an audit.
      a. True
      b. *False*

3. Self-management involves
      a. Determining the three or four things that are most important to you.
      b. Making time to reflect on your personal needs and goals.
      c. Making time to reflect on your current progress and obstacles.
      d. *All of the above.*

4. When preparing a meeting agenda, less important issues should
      a. Be excluded because the risk is insignificant.
      b. *Go at the top of the agenda so that they don't get eclipsed by the bigger items.*
      c. Be sent in an e-mail after the meeting.

CHAPTER 7: TECHNIQUES FOR MANAGING THE AUDIT TEAM

1. Effective teams (circle all that apply)
      a. Share common values and goals.
      b. Have strong, top-down leadership.
      c. *Follow a consistent and logical methodology.*
      d. All of the above.

2. The acronym TACT stands for
      a. Tactful Always Considering Ties.
      b. *Tell Affect Change Trade-off.*
      c. Talk About Change Timely.

3. You should provide feedback as soon as an undesirable behavior occurs to prevent repetition.
   a. True—this avoids repetition of the behavior and quick correction.
   b. *False—individuals need time to self-assess and self-correct.*
4. It is better to be a leader than a manager.
   a. True—in general, leaders are forward-thinking and managers maintain the status quo
   b. *False—both are needed in a successful organization.*
5. When leading a multigenerational team, focus on
   a. Developing skills in those of your generation first.
   b. *Teaming different generations to leverage strengths.*
   c. Accommodating each team's preferences as much as possible.

CHAPTER 8: TECHNIQUES FOR MANAGING THE CONSTITUENT RELATIONSHIP

1. It's right to "judge a book by its cover."
   a. Never, we must give people time to display themselves.
   b. *It doesn't matter if it's right—it happens, and we need to make sure we reached the right conclusion.*
   c. Absolutely—what you see is what you get.
2. Since you are the auditor, it is your responsibility to establish the relationship
   a. *True—I am initiating the project, and I can set the tone for the next few months.*
   b. False—audit requires independence, not the ability to influence.
3. It's important to talk to the people about something other than the audit.
   a. *True—this leads to mutual trust and a camaraderie based on real feelings.*
   b. False—work is work—everything should stay in its place to avoid blurry boundaries.
4. The most important aspect of facilitating is
   a. Setting the foundation by creating a proper agenda.
   b. *Focusing on how to bring together of diverse backgrounds, interests, and capabilities.*
   c. Making sure you are the best prepared.
   d. None—they are all important.
5. Your communication approach should stay the same at all points of the audit.
   a. True—the client and your team needs to feel that you are consistent.
   b. *False—different stages of the audit may require your communication style to change.*

Bibliography

Cohen, M. A., Rogelberg, S. G., Allen, J. A., and Luong, A. 2011. Meeting design characteristics and attendee perceptions of staff/team meeting quality. *Psychology Faculty Publications*. Paper 96.

Doran, G. T. 1981. There's a S.M.A.R.T. way to write management's goals and objectives. *Management Review* (AMA FORUM) 70 (11): 35–36.

Elder, R. J., Akresh, A.D., Glover, S. M., Higgs, J. L., and Liljegren, J. 2013. Audit sampling research: A synthesis and implications for future research. *Auditing: A Journal of Practice and Theory* 32, 99–129. *Business Source Premier*, EBSCO*host* (accessed June 25, 2015).

Hall, T. W., Higson, A. W., Pierce, B. J., Price, K. H., and Skousen, C. J. 2013. Haphazard sampling: Selection biases and the estimation consequences of these biases. *Current Issues In Auditing* 7 (2): 16–22. *Business Source Premier*, EBSCO*host* (accessed June 25, 2015).

Hitzig, N. B. 2004. Statistical sampling revisited. *The CPA Journal*, 30. *Academic OneFile*, EBSCO*host* (accessed June 25, 2015).

Sibelman, H. 2014. Myths and inconvenient truths about audit sampling: An audit partner's perspective. *The CPA Journal*, 6. *Academic OneFile*, EBSCO*host* (accessed June 25, 2015).

Index